Make:
Getting
Started with
Drones

Terry Kilby and Belinda Kilby

MAKER MEDIA™
SAN FRANCISCO, CA

Make: Getting Started with Drones

by Terry Kilby and Belinda Kilby

Published by Maker Media, Inc., 1160 Battery Street East, Suite 125, San Francisco, CA 94111.

Maker Media books may be purchased for educational, business, or sales promotional use. Online editions are also available for most titles (http://safaribooksonline.com). For more information, contact our corporate/institutional sales department: 800-998-9938 or corporate@oreilly.com.

Editor: Patrick Di Justo
Production Editor: Matthew Hacker
Copyeditor: Eileen Cohen
Proofreader: Jasmine Kwityn

Indexer: Angela Howard
Interior Designer: David Futato
Cover Designer: Brian Jepson
Illustrator: Rebecca Demarest

October 2015:   First Edition

Revision History for the First Edition

2015-10-02   First Release

See http://oreilly.com/catalog/errata.csp?isbn=9781457183300 for release details.

978-1-457-18330-0

[LSI]

Contents

Preface

We are Belinda and Terry Kilby. We are drone enthusiasts, aerial photographers, makers, trainers, and a husband–wife team. By combining lives, we also combined our strengths and passions for technology and art. Since 2010, we have been designing and building small unmanned aerial vehicles (UAVs) for artistic and practical aerial photography through our company, Elevated Element. As early adopters, we became the unofficial spokespeople for UAV technology for the media in our region. We keep up with the advancements in hardware, software, and drone news stories, so we can represent drone builders and users in an artistic and innovative light.

The First Book of Its Kind

In the fall of 2013, we released *Drone Art: Baltimore*, the first photo book shot entirely using small, custom-built UAVs. The introduction describes how we got started and how our equipment and work evolved. The images are presented in chronological order to show progression in quality, as we came to understand how to build, fly, and take photographs with these drones. The book launch, which we arranged through Baltimore's Office of Promotions and the Arts, was held at an exhibition at the Baltimore World Trade Center, on the observation deck called "Top of the World." The location was perfect, because the perspective looking out of the windows from the 27th floor was much like the subject matter seen in our work: the bird's-eye view.

The Book's Goal

Our objective for this book is to give you that same working knowledge of aerial robotics, by showing you how to build a Little Dipper 300-class autonomous quadcopter as an example. By reading and following the steps and advice contained in this book, you'll learn how quadcopters work and how to solve some of the engineering challenges a quadcopter presents. Where applicable, we will also suggest alternative options you can try, or comparable parts to suit your own preferences. Whether you choose to build a quadcopter or an octocopter, the same concepts will apply.

The Little Dipper design is open source, and the design files can be downloaded from *http://gettingstartedwithdrones.com/little-dipper-build/*. Alternatively, you can order a complete Little Dipper kit, with all the hardware, from *www.MakerShed.com*.

And don't worry about failure. Everyone makes mistakes. It's only important that we continue to try, do, and make. Aerial robotics is a hands-on experience, in which you solve real-world problems through trial and error, and ultimately, hard-won wisdom. The smallest difference in weight distribution or speed can mean success or disaster.

How to Use This Book

Each chapter of this book is designed to cover a specific portion of the overall technology of unmanned aircraft. One chapter covers the intricacies of different airframe types, while another dives into how the Global Positioning System (GPS) is used to assist flight. We made a conscious effort to touch on as many topics as we could, but keep in mind that this is a Getting Started with book, not the be-all-and-end-all encyclopedia of UAVs. If you are a newcomer to this technology, there will be plenty of meat here for you to chew on. If you are already familiar with drones, we are sure there will still be some important nuggets of info that you can add to your knowledge base.

Once you get past Chapter 1, you should start to see an important pattern emerge: most chapters are broken down into a section for background and theory followed by step-by-step instructions for our build example. No matter if you are building the same aircraft we show throughout the book, the theory portion of each chapter will apply to almost any type of drone— from a small 250-class quadcopter to a huge 1,000-class octocopter. If you are not following along with our demo build, feel free to skip over the build instructions and consume as much of the theory as you can.

Whether you are a seasoned drone pilot or just getting into the hobby, we think that everyone can benefit from the theory portions of each chapter. We do recommend that you read each of them. If you are also following along with the example build, feel free to either build as you go or read through all the theory first before finally coming back and starting on the build. A thorough understanding of the technology involved can only help you when it comes time to build your aircraft.

Conventions Used in This Book

The following typographical conventions are used in this book:

Italic
    Indicates new terms, URLs, and email addresses.

This element signifies a general note, tip or suggestion.

This element indicates a warning or caution.

Safari® Books Online

Safari Books Online is an on-demand digital library that delivers expert content in both book and video form from the world's leading authors in technology and business.

Technology professionals, software developers, web designers, and business and creative professionals use Safari Books Online as their primary resource for research, problem solving, learning, and certification training.

Safari Books Online offers a range of plans and pricing for enterprise, government, education, and individuals.

Members have access to thousands of books, training videos, and prepublication manuscripts in one fully searchable database from publishers like Maker Media, O'Reilly Media, Prentice

Hall Professional, Addison-Wesley Professional, Microsoft Press, Sams, Que, Peachpit Press, Focal Press, Cisco Press, John Wiley & Sons, Syngress, Morgan Kaufmann, IBM Redbooks, Packt, Adobe Press, FT Press, Apress, Manning, New Riders, McGraw-Hill, Jones & Bartlett, Course Technology, and hundreds more. For more information about Safari Books Online, please visit us online.

How to Contact Us

Please address comments and questions concerning this book to the publisher:

Make:
1160 Battery Street East, Suite 125
San Francisco, CA 94111
877-306-6253 (in the United States or Canada)
707-639-1355 (international or local)

We have a web page for this book, where we list errata, examples, and any additional information. You can access this page at *http://bit.ly/gs_w_drones*.

To comment or ask technical questions about this book, send email to *bookquestions@oreilly.com*.

1/Introduction

Some Definitions

Unless you've been living under a rock, you're probably aware that the word *drone* is frequently in the news. The many headlines about drones have used the term to describe a wide range of aircraft—from small remotely piloted toys, to autonomous flying robots, to full-scale weaponized military surveillance models. This is mainly because different sources have had different definitions of the word *drone*. Where exactly is the line drawn, or what makes a drone a drone? Let's start with a basic definition.

Merriam-Webster's definition of *drone* is:

> an unmanned aircraft or ship guided by remote control or onboard computers

This definition presents a very broad sense of the word, which contributes to the overgeneralizations and misinformation we see when the media reports on a particular type of unmanned aircraft. Let's be more specific. Terry says he draws the line between radio-controlled (RC) aircraft and drones at the introduction of GPS and autopilots. When an aircraft has the ability to pilot itself, even if it's just to hold a steady position, that in his eyes is a drone. Throughout this book, we'll use the following conventions:

Drone
> Unmanned aerial vehicle controlled autonomously using GPS

Remotely piloted aircraft (RPA)
> Model aircraft flown by a pilot on the ground using a radio transmitter or other computer equipment

UAV
> Aircraft that can be flown remotely by a pilot or controlled autonomously using computer software and GPS

Small Unmanned Aerial Systems (sUAS)
All related processes involved with unmanned aerial technology

Whether we like it or not, the word *drone* will continue to be used in a sweeping fashion. We are embracing the word and are aiming to help change the negative connotations, by showing the beneficial applications of small UAV technology. With that in mind, we need to increase our own understanding to navigate the media frenzy that is drone-mania. While the US Federal Aviation Administration is trying to settle on comprehensive small UAV commercial use policy, we all need to take our aerial pursuits as responsibly and safely as possible.

Who Is This Book For?

This book is a set of instructions (with additional suggestions along the way) for how to build an autonomous quadcopter. A general understanding of robotics and electronics concepts are a real advantage in pursuing aerial robotics. It also helps to be familiar with basic tools and equipment, including a soldering iron, to have long-term success in designing your own small UAVs.

If you're a maker who enjoys persevering through trial-and-error problem solving while you're building something, then you'll enjoy aerial robotics. Being able to build a flying robot, and view scenes from completely new perspectives, is well worth the time and effort.

The Drone User Community

Sometimes the way to success when you have a problem is knowing the right questions to ask others. Networking with people who share an interest in aerial robotics is an invaluable resource for helping to pinpoint issues and finding solutions. Online forums are a wonderful way to see how others solve similar problems. One site that has been a favorite of Terry's is Multi-RotorForums.com (*http://www.multirotorforums.com*). People on that forum have been incredibly generous in sharing their experience and insights in building and flying small UAVs.

There may also be an organized group of UAV enthusiasts or a model airplane club local to you, either of which would probably appreciate seeing a fresh face at a meeting. Try searching Meetup.com (*http://www.meetup.com*) for drone (there's that word again!) user groups. Another place to look is the Academy of Model Aeronautics (*http://www.modelaircraft.org*). The AMA is the world's largest model aviation association and has been around since 1936. We have a great group of folks here in the Baltimore and Washington, D.C., area who have contributed greatly to sharing UAV technology in our region. We truly appreciate everyone who has helped us maintain this wild endeavor.

There may also be an organized group of UAV enthusiasts or a model airplane club local to you, either of which would probably appreciate seeing a fresh face at a meeting. Try searching Meetup.com (*http://www.meetup.com*) for drone (there's that word again!) user groups. Another place to look is the Academy of Model Aeronautics (*http://www.modelaircraft.org*). The AMA is the world's largest model aviation association and has been around since 1936. We have a great group of folks here in the Baltimore and Washington, D.C. area who have contributed greatly to sharing UAV technology in our region. We truly appreciate everyone who has helped us maintain this wild endeavor.

The AMA has a great PDF of best practice guidelines (*http://bit.ly/ama_afsc_guidelines*) for safe and responsible flight.

Brief History of Autonomous Flight

The top inventions that we believe have contributed most to drone technology include the RC model airplane, microchips, GPS, the Internet, and the smartphone. Let's take a look.

RC Model Airplane

In 1937, Ross Hull and Clinton DeSoto, officers of the American Radio Relay League, performed the first public demonstration of remote-controlled flights. In the summer and fall of 1937, they designed and built sailplanes with a 13-foot wingspan, completing over 100 successful radio-controlled flights in Hartford, Connecticut. During this era, Hull set the pace for homebrew radio apparatus design. He increased transmitter efficiency by shortening the leads and was the first to describe the much lighter, one-tube on-board receiver for model aircraft. Twin brothers Walter and William Good won first place titles in 1940 and 1947 at the US National Aeromodeling Championships. Their iconic RC model airplane, known as the Guff, is now owned by the Smithsonian National Air and Space Museum (see Figure 1-1).

Figure 1-1. *The Good brothers' RC airplane, the Guff.*

The Advent of Microchips

In the summer of 1958, Jack Kilby—a new employee at Texas Instruments and young inventor at the time—revolutionized the electronics industry with the introduction of his integrated circuit. This precursor to the microchip consisted of a transistor and other components on a thin piece of germanium 7/16 × 1/16 inches in size. Knowing that many electronic components, like passive resistors and capacitors, could be made from the same material as the active transistors, Kilby realized they could also be made into configurations to form a complete circuit. Many electronics we use now would not be possible without Kilby's tiny chip. It transformed room-sized computers into the microcomputers sold today.

The Technology of Drones

At a certain point, model aeronautics reached the maximum height that hardware design, radio signals, and electronic pulses could take it. To go beyond would require the implementation of less tangible technologies that would enable intelligent communication and control.

The Launch of GPS

The official GPS.gov site (*http://www.gps.gov/*) describes GPS as follows:

> The Global Positioning System (GPS) is a U.S.-owned utility that provides users with positioning, navigation, and timing (PNT) services. This system consists of three segments: the space segment, the control segment, and the user segment. The U.S. Air Force develops, maintains, and operates the space and control segments.

The system's 36 satellites constantly broadcast a stream of time-code and geographical data to users on the ground. Any device with a GPS receiver can use data from any four satellites to calculate its location in relation to those satellites. Maintaining a clear line of site with the GPS satellites is key, and accuracy is improved as you connect to more than the minimum of four. Due to the line-of-site requirement, it can sometimes be difficult to obtain a reliable GPS lock indoors. We will discuss how this affects flying a drone indoors in Chapter 5.

 More GPS Information

For additional details, see the following pages from GPS.gov:

- "How GPS Works" (*http://bit.ly/gps_ed_poster*)
- "GPS Applications" (*http://bit.ly/gps_apps*)

Internet

The personal, civilian drone boom would not be where it is today without the Internet. Online shops, social media, and forums enable people to instantly share and learn with people anywhere in the world. Terry got through his initial quadcopter builds by observing other people's designs and asking questions online. The more intelligent the UAV, the more of a role the Internet will play in future drone applications.

The Smartphone

With the ability to dramatically shrink the size of computer processors and sensors, it was only a matter of time before someone had the idea to use a smartphone's insides on a model aircraft. When you turn or rotate your smartphone, the orientation of the interface changes direction; the same sensors could be used to control a small drone. As a mobile software engineer, Terry was familiar with developing mobile apps and the capabilities of smartphone operating systems. Currently, he is working on a number of different applications in the drone mapping space.

Small Autopilot Flight Controller

All these things—GPS, the Internet, and the smartphone—have led to the flight controller, essentially the brain of the drone. Civilian autopilots started showing up on hobbyist multicopters around the late 2000s. Early GPS-capable units were available from the German company MikroKopter, and then several Chinese companies copied them. Around that same time, several open source projects started up, such as MultiWii (*http://www.multiwii.com/*), Ardupilot (*http://ardupilot.com/*), and Open Pilot (*https://www.openpilot.org/*). MultiWii took its name from the interesting fact that the first units were made with sensors hacked from a Nintendo Wii controller. Ardupilot, as you can probably guess, was so called because it was originally Arduino based.

Today, the small autopilots have came a long way, and many include advanced features such as autonomous flight, Return to Home, and Follow Me. Many of these features were only available on top-of-the-line models just a few years ago, but that goes to show you how fast this technology is evolving.

Small Autopilot Sensors Needed for Flight Control

The following sensors, while not brand new, have finally become small and light enough to allow their use in UAV autopilots:

Magnetometer
   Digital compass

Gyroscope
   Measures rates of rotation

Accelerometer
   Measures gravity

Pressure sensor
   Calculates altitude, by measuring atmospheric pressure

Combined, these sensors create an Inertial Measurement Unit (IMU).

Principles of Flight

The mechanics of flight consist of some simple rules with complex interactions. To understand them well, it wouldn't hurt to spend some time brushing up on Newton's laws of physics.

When we talk about a *force*, what we mean is a simple push or pull. If the forces working on an object are balanced—a push in one direction met by an equal push in the opposite direction—the object is stationary. If the forces are not balanced, the object accelerates in the direction of the stronger force.

Weight/Gravity

Weight is the force on an object caused by *gravity*. The principle of force is also sometimes called gravity. For something to fly, or even hover, it must somehow continuously balance or overcome the force of gravity (we'll see how it does that in a moment).

Gravity is relentless—even a momentary loss of the opposing force can bring the aircraft crashing to earth. One interesting note about gravity we'll be dealing with throughout the book: although weight is distributed throughout the aircraft, one point in the aircraft—called the center of gravity—has the most effect on its ability to fly.

Lift

Lift, the opposite of weight, is an aerodynamic force that keeps an aircraft in the air (see Figure 1-2). In the case of winged aircraft, lift comes from air moving across an airfoil shape of a wing or propeller. The air moving above the airfoil is moving faster; therefore, it has lower pressure. Slower-moving air below the wing has higher pressure. Thanks to the lower pressure above the wing, an airplane or helicopter is literally sucked into the sky. To hover or fly level, lift must equal weight; to climb, lift must be greater than weight.

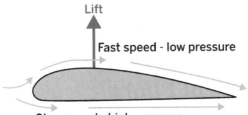

Figure 1-2. *As the airfoil form moves forward through the air, it produces lift.*

Drag

Have you ever stuck your hand out the window of a moving car on a nice day? The force you felt pushing back is a perfect example of *drag*. Any object that moves through the atmosphere at any speed will experience some level of drag, and it increases with the speed of the object. Drag is the reason airplanes, locomotives, and sports cars have smooth, sleek lines—that type of streamlining allows air to flow more cleanly around the vehicle, cutting down on drag and making the vehicle more efficient. Drag is also the reason jets retract their landing gear right after takeoff, and it can be a potent force for quadcopters/drones.

Thrust

The *thrust* principle of flight is the mechanical force that moves an aircraft through the air. The motion must be created in some way by engines, propellers, rockets, muscles (in the case of birds that can fly), or whatever propulsion system is employed. If thrust is greater than drag, the aircraft will increase in speed. Thrust must equal weight and overcome drag.

Figure 1-3 illustrates the four principles or forces of flight.

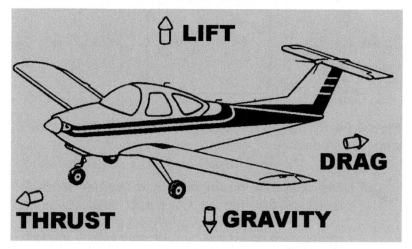

Figure 1-3. *Aircraft illustrating the principles or forces of flight.*

Flight Maneuvers: Aircraft Movement with Stick Mapping

Most UAVs are controlled with a standard six-channel (minimum) remote control just like those that model airplanes have used for many years. The remotes have two main joysticks that move both forward and backward as well as left to right. You will also see some combination of switches, knobs, and sliders, depending on your model. Each of these input mechanisms occupies one channel of your radio.

The two main sticks are the most important controllers and occupy four channels total, one for each axis the stick can travel. Outside of the main sticks, we will always need at least one channel to control the aircraft's flight mode setting. Another common channel requirement is for a feature called Return to Home (RTH). Both of those channels are assigned to and controlled by a switch. We will discuss RTH and flight modes in greater detail later in the book; for now, let's look more deeply at the channels controlled by your two main sticks (see Figures 1-4 and 1-5).

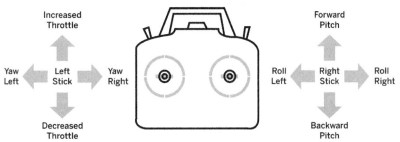

Figure 1-4. *RC transmitter left and right stick command movements for Mode 2.*

The controller descriptions here are for Mode 2 radios used in the United States. Other countries may use Mode 1 radios, which simply reverses the movements controlled by each stick. Many modern radios can use either mode, but some are configured one way only at the factory when they are built.

Throttle

Forward/backward motion of the left stick controls the throttle of your aircraft. The throttle essentially acts as the gas pedal for your aircraft, just as its name implies. In most cases, the higher the throttle value, the faster your motors will spin. Of course there are exceptions to that, and we will talk about those in reference to autopilot and autonomous flight. In order for your aircraft to hover above the ground, the throttle must generate enough lift to counter the effects of *Weight*. During forward flight, the throttle must counter both drag and weight.

Yaw/Rudder

Left-right motion of the left stick is the channel for *yaw*, which can also be called *rudder*. Yaw controls rotation across the horizontal axis of a helicopter or multirotor. If you are flying a plane, this channel would be called rudder due to its control of the tail flap by the same name. The effect on flight of the aircraft is the same for both yaw and rudder during forward flight: steering the aircraft in the required direction.

How does any multirotor, with only propellers for moving parts, mimic flight maneuvers that a plane needs flaps and a rudder to achieve? It's all done through a process called *vector thrusting*. We will get into it more in the next chapter, but the basic idea is to control the speed of each propeller independently in such a manner to move in any direction. You could, for example, yaw clockwise by increasing the speed of the two clockwise propellers and decreasing the speed of the two counterclockwise props.

Pitch

Forward-backward motion of the right stick is the channel for *pitch*, which is also referred to as *elevator*. Pitch tilts the nose of the aircraft up or down. When you move the right stick forward, the nose of the aircraft will pitch down and vice versa. On an airplane, this is achieved by tilting the horizontal tail flaps together in the same direction. Your quadcopter will be able to move in the same manner by the use of vector thrusting, as with yaw. Most autopilots have an autolevel mode that places a limit on how far you can pitch an aircraft. Other modes may not place this limit, allowing the aircraft unlimited pitch. Under the proper conditions, it is even possible to pitch all the way into a forward flip, but you'd better do a little practicing before trying that!

Roll

Left-right motion of the right stick is the channel for *roll*, also known as *aileron*. Roll tilts the aircraft to either the left or right in relation to the front of the aircraft. With an airplane, this would happen by tilting the horizontal wing flaps (known as the ailerons), in opposite directions from each other. Vector thrusting is also responsible for handling all roll movement. This attitude change causes the aircraft to fly in the direction of the tilt. Just like pitch, maximum roll is capped off in autolevel modes and is unlimited in manual mode. With the right autopilot settings and a little practice, a small aircraft can do barrel rolls just like a plane.

Practice Makes Perfect: Using a Flight Simulator

Begin practicing flight-stick commands as soon as possible using flight-simulator software (*http://gettingstartedwith drones.com/simulators/*) on a computer. We recommend that you research to find one you feel comfortable with. We used the Phoenix Professional Flight Simulator in our Quadcopter Maker Camps. It is a software disk that comes with an actual remote control that hooks up to your computer using a USB cable. There are also several mobile apps, including a couple of free ones, that you can download. Another simulator that we currently use is Heli-X. We enjoy it, because it works on the Mac. The simulator is a valuable tool for beginner pilots to build experience and confidence without wrecking expensive equipment. More experienced operators can also use it to keep their skills sharp. Being able to practice flying no matter what the weather is like outdoors is great. Terry spent an entire winter flying with simulator software to build his skills early on. Strong muscle memory of stick commands makes all the difference if you encounter a challenge while piloting.

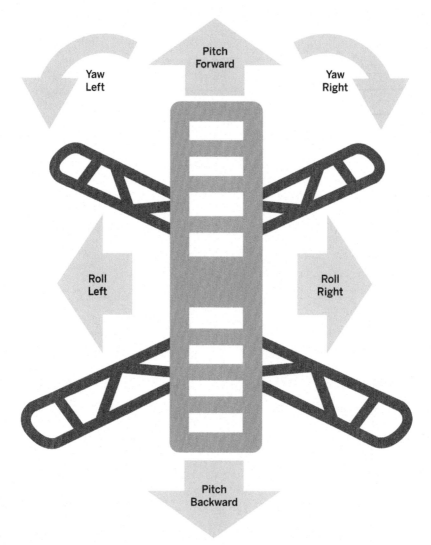

Figure 1-5. *Basic quadcopter showing how the stick commands would move the craft.*

2/Airframes

What Is an Airframe?

The airframe is the main body of the aircraft. All other components—propellers, batteries, computer, etc.—are mounted to the airframe. Airframes can vary greatly in size and complexity. Some of the airframes we built when we first got interested in flying were very simple designs sketched on the back of a napkin, and cut from pressboard with a jigsaw. Today, we use computer-aided design (CAD) software to design our frames before cutting them out with a computerized numerical control (CNC) machine. We will touch on many of the popular design types in this chapter, but first let's take a quick look at how these actually fly.

Thrust Vectoring

One common trait you will see with all of these airframes is that they use both clockwise and counterclockwise propellers. This is the foundation that allows multirotors to move in all directions. Every other motor spins in a different direction from the one directly next to it, allowing each motor to be balanced by its opposite motor. As a result, nearly all UAVs feature an even numbers of propellers (with one exception that we will discuss in the next section). On a quadcopter, for example, the NW and SE motors might spin clockwise while the NE and SW motors spin counter-clockwise (speaking in terms of compass bearings). By configuring the aircraft this way, all we have to do to move the quadcopter in any direction, including yaw, is change the speed of a certain combination of motors. For example, to yaw the aircraft in a clockwise direction, we would increase the speed of the NW and SE motors. The "push" of one motor and the "pull" of the other will turn the aircraft in the direction we want.

Why Opposite Directions?

You may have noticed that our propellers spin in opposing directions from one to the next. There is a very good reason for this. That process provides counteracting balance to the aircraft. If you used propellers that all spun in one direction, the aircraft would naturally want to spin in that direction, and the flight controller would have to fight that tendency at all times.

Another way to think about it is by examining a full-size helicopter. The main rotors that spin above the aircraft obviously spin in only one direction. The tail rotor provides the counterbalance that keeps the aircraft heading straight. If you removed the aircraft from the equation, the helicopter would still fly, but it would spin around the rotor axis the entire time it was airborne.

Aircraft Designs

Figure 2-1 shows the various frame types we'll look at in the following sections. Notice the direction that each prop rotates compared to its neighbor.

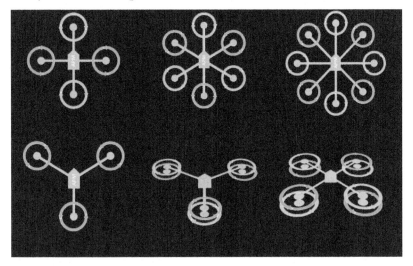

Figure 2-1. *Frame types showing the directional rotation of the propellers.*

Tricopter

The tricopter is the only common multirotor that doesn't have an even number of props. It achieves yaw movement by placing one of the three motor/prop combos (typically the one in the back) on an axle that allows for angled thrust. It features three propellers total.

Quadcopter

Probably the most popular design today, the quadcopter offers the simplest mechanical design with the least number of required components. It features four propellers total.

Hexacopter

Another very popular design is the hexacopter, due to its increased ability to carry a payload while still remaining fairly agile. It features six propellers; you may also hear this design called a "flat six."

Octocopter

The octocopter design is often used for aircraft that need increased payload along with redundancy. Because it features a total of eight propellers, it is possible for an octocopter to remain in flight if there is any type of failure with any one of the motors or props. The remaining seven will keep the craft in the air without incident. You may also hear this design called a "flat eight."

Y6

The first of our "coaxial" aircraft, the Y6 features six motors and propellers installed on an airframe with three arms. This is accomplished by stacking motors and props on the top and bottom of each arm with each spinning in the opposite direction from the other. You have probably seen coaxial designs on common store-bought RC helicopters. This design type is inherently more stable but 20%–25% less efficient than a traditional "flat" design.

X8

Another popular coaxial design is the X8. The frame itself looks almost identical to a quadcopter, with the main difference being that each arm now has two motors and props for a total of eight.

Materials

Many different types of materials are used to build the modern-day drone. Some of the more popular options include carbon fiber, fiberglass, and various types of plastic or metal. As with most things in this hobby, you'll always perform a balancing act based on which characteristics are important to you. There is an old saying in this hobby, "Cheap, strong, or lightweight: you can have any two." While materials like carbon fiber may have a very high weight-to-strength ratio, they can be very expensive to work with. For this reason, many people begin by making their designs from wood, or even 3D-printed acrylonitrile butadiene styrene (ABS) plastic. The frame that we will be working through in this book is called the Little Dipper and is primarily made from G-10 fiberglass (see Figure 2-2). We think this material is a nice balance of weight, strength, and cost.

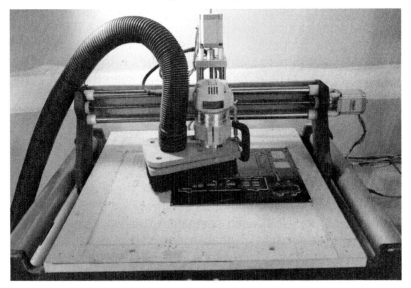

Figure 2-2. *G-10 fiberglass frames being cut on a CNC machine.*

Keeping It Balanced

No matter what type of frame you are building, it's always very important to keep the weight balanced front to back and side to side. If you are designing something from scratch, keeping your frame symmetrical is a great way to ensure that it will be easy to balance after your build is complete. If you have to go with a slightly asymmetrical design, try to position your components on the frame in a way that ultimately brings it back into balance.

Your drone's battery is often a good candidate to reposition in order to bring balance to your design.

Building the Little Dipper Airframe

The Little Dipper is a compact, foldable quadcopter. Its airframe is made up of two subframes that help isolate the motor vibrations from the flight and imaging sensors. These subframes are called the clean and dirty frames. The dirty frame is the bottom subframe, and it holds all of the moving parts such as the motors and the propellers. The clean frame sits on top and holds all of the flight and communication electronics.

Lay out the flat frame parts, and refer to the photographic diagram in Figure 2-3 to identify them. Divide them into the two subframe build portions. Parts A and B make up the dirty frame (A is bottom, B is top), while C identifies the four booms, which will hold our motors and props. Parts D, E, and F make up the clean frame: D identifies the bottom clean frame plate, while E shows the top clean frame plate. Part F is the camera plate, which sits on top of rubber isolation balls (more on that soon).

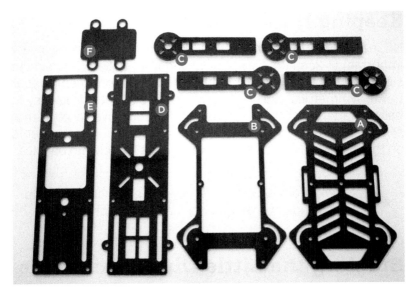

Figure 2-3. *You should have all of these parts in your frame kit.*

What Makes It Clean

The terms *clean* and *dirty* might seem like a strange way to describe frames. We aren't talking about the part of the frame that sits closest to the ground and therefore is most likely to get dirty. We are talking about vibrations. The theory behind these types of subframes is simple: build one subframe to hold all the moving parts—which cause vibrations—and another to hold everything else. We isolate these two subframes by providing the least amount of contact possible between them. For the Little Dipper, we ship it with four small aluminum standoffs that mount the two frames together. It is possible to isolate them even further by replacing the aluminum standoffs with rubber versions. During our testing, we found very little difference in performance between the two. As a result, we believe that the aluminum standoffs last longer, and that is ultimately what drove us to including them in the frame kit.

Alternative Frame Options

Although we are walking you through our example build with the Little Dipper, a few alternative options for airframes are presented in Table 2-1.

Table 2-1. *Airframe alternatives to the Little Dipper*

Frame Name	Configuration	Independent Booms	Clean/ Dirty Frame	Folding Booms
Lumenier QAV250	250-class quadcopter	No	No	No
Blackout Mini Spider Hex	290-class hexacopter	Yes	No	No
DroneKraft Mach 300	300-class quadcopter	Yes	Yes	No

The QAV250 (see Figure 2-4) is one of the most popular frames on the market today. It's extremely lightweight due to its simple two-plate design and very effective for quadcopter racing.

Figure 2-4. *QAV250 with 5X3 carbon fiber props.*

Step-by-Step Build Instructions

Here are the tools you will need to get started (see Figure 2-5):

- Metric Allen wrench (2.5 mm). A full set is nice to have, but we will be using the 2.5-mm wrench during this build.
- An adjustable wrench, needle-nose pliers or a 7/32 nut driver. Any of these will work, but if you use the nut driver, make sure it is slim enough to fit into the tight spaces around the frame. A full ratcheting socket will not work, because it is too large. The needle-nose pliers will mark up your hardware slightly, so use those as a last resort.

Figure 2-5. *Tools you will need for the frame build.*

Step 1: Install the Standoffs

Begin building the bottom portion of the airframe—the dirty frame—by attaching the short aluminum standoffs (see Figure 2-6). Locate the top dirty frame plate (plate B in Figure 2-3). There are four 3-mm holes at each corner of the plate (see Figure 2-7). Use those holes to mount the standoffs with the 5-mm screws (the short black screws included in the kit).

Figure 2-6. *Standoffs—either aluminum for stiffness and longer life, or rubber for increased vibration isolation—are used to separate the two subframes.*

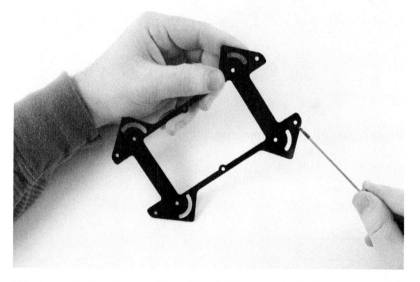

Figure 2-7. *Pointing out the standoff mounting holes on plate B of the dirty frame.*

Manually fit the screws through the holes, then attach the standoff on the other side of the plate. The screws should not come out the other side of the standoff (if that happens, you used the wrong screws). When all four are done, plate B should look like an upside-down table with its legs sticking up in the air, as shown in Figure 2-8.

Figure 2-8. *All four aluminum standoffs installed on plate B.*

Which Way Am I Facing?

In Figure 2-8, plate B is pointing away from the camera and right side up. That is to say, the front of the drone—and this single plate—is the end at the top of Figure 2-8. There are two clear indicators that tell us this: first, the short standoffs are pointing up; second, the arched boom mount holes (more on those in a minute) are swiveling toward the *back* of the aircraft. You can also identify the front of the frame because it has longer (or wider, depending on how you look at it) arched boom holes. Keep these indicators in mind as you move through the build.

DroneKraft Mach300

If you are building out a Mach300, your kit will come with rubber standoffs very similar to those shown in Figure 2-6 and should provide a comfortable level of vibration isolation for your flight controller.

Step 2: Complete the Dirty Frame Assembly

Complete the bottom portion, or dirty frame assembly, by attaching the four booms (plates C in Figure 2-3) and bottom dirty frame plate (plate A) to your top frame plate. When you are finished, it will look like Figure 2-9.

Figure 2-9. *The dirty frame fully assembled from plates A, B, and C.*

It is best to do this step one boom at a time. Begin by placing plates A and B on top of each other with the standoffs on plate B pointing up. Make sure that both plates are facing the same direction. You can check this by locating the boom holes (shown in Figure 2-10) and making sure that the arched half-moon holes line up between both plates. If they don't, flip one of the plates (front to back) until the holes match up.

Why Are Those Holes Arched?

You may notice that the boom mounting holes shown in Figure 2-10 are made up of one normal round 3-mm hole and one arched slot. There is a simple reason for this design. The combination allows our booms to fold back and allow for a more compact design as well as to soften the impact in the event of a crash. Folding arms will give way if you crash with any type of forward motion, whereas solid arms are much more susceptible to breaking.

You can identify the boom holes in plates A and B pretty easily. Both front and back mounting holes will have a single 3-mm round hole positioned just above an arched 3-mm slot that allows folding of the frame.

Figure 2-10. *Rear boom mounting holes (notice the anchor hole just above the folding slot).*

Let's continue by sandwiching the boom (plate C in Figure 2-3) between the top and bottom dirty frame plates (plates A and B); see Figure 2-11.

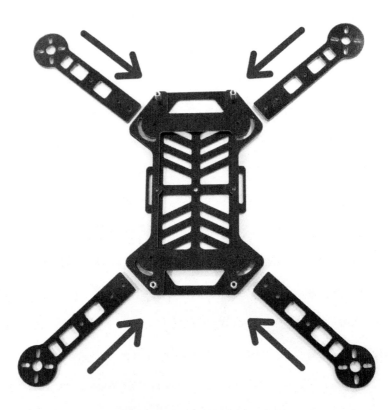

Figure 2-11. *The booms are lined up and ready to be sandwiched between the top and bottom dirty frame plates.*

 Stand Up Straight

The standoffs are added to the design to help isolate vibrations between the clean frame and the dirty frames, so they should be on the *top* of the dirty frame top plate (plate B).

Now take your first boom and line the two 3-mm holes on the rectangular end with the boom mounting holes. Take a 12-mm mounting screw (the longs screws that came in the kit) and place a flat black washer over it. Now manually push this screw and washer through one of the boom mounting holes in the top plate, then the boom, and finally the bottom plate. We recommend starting with the 3-mm anchor hole first. Once the screw is all the way through the sandwiched set of plates, push another flat black washer on the other side of it before finally placing a 3-mm locking nut on the bottom. Repeat this for the second hole on the boom (see Figures 2-12 and 2-13). Once you have one boom installed, repeat this entire step for the rest of the booms.

Figure 2-12. *Pushing the second screw through the mounting holes.*

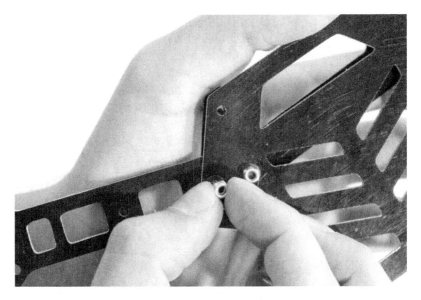

Figure 2-13. *Placing the nuts on the bottom of the frame.*

Do Not Overtighten the Nuts

You can very easily overtighten the nuts on the boom mounting screws. This will not cause any damage, but it will prevent your booms from folding. You want to tighten them just enough to hold the boom in place firmly, but not so much that it prevents it from folding at all. Practice folding the boom as you are going through this step. The nuts we use, called lock nuts, have a layer of nylon at the end of the thread that holds the nut in place without having to overtighten it. You should still consider checking the tightness of the mounting screws/nuts during regular aircraft inspections.

If you are building a Blackout Spider Hex frame, your booms will mount to the bottom plate in a very similar fashion. The main difference you will see is that you use four screws instead of two and the booms will not fold into the body.

Step 3: Assemble the Clean Frame

Next, we are going to build out our clean frame. This frame is made up of plates D, E, and F (which we will add later in the book) from Figure 2-3. We will also be using the eight 37-mm standoffs that came in the kit. The overarching idea is to use plates D and E as a top and bottom of our frame that is held together with the standoffs.

Let's start by finding plate D. You will notice eight 3-mm diameter holes that are positioned around the edges of the plate. These are our standoff mounting holes. Take a 5-mm black screw (one of the short screws that came with the kit) and manually push it through one of the holes (see Figure 2-14). It doesn't matter which hole you pick, because we will repeat this for all the holes. You also do not need a washer for these screws.

Next, take one of your 37-mm standoffs and thread it on the portion of the screw that is sticking out of the other side of plate D (see Figures 2-15 and 2-16). Once you have it finger-tight, you can tighten it further with your Allen wrench and adjustable wrench or pliers.

Figure 2-14. *Pushing the 5-mm screw through the standoff mounting holes on plate D.*

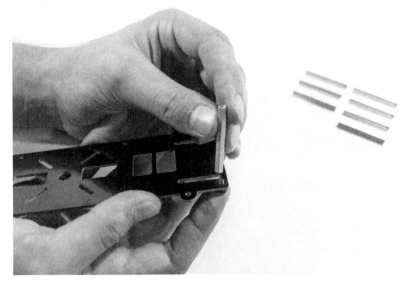

Figure 2-15. *Threading the first 37-mm standoff on the back of our mounting screw.*

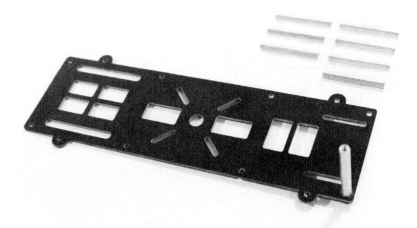

Figure 2-16. *Our first standoff is installed (the rest are waiting in the wings).*

Now we can repeat these steps for the remaining 37-mm standoff (see Figure 2-17). When you are all done, plate D should like Figure 2-18.

Now that we have our 37-mm standoffs installed on plate D, we are ready to attach plate E. If you haven't done so already, find that plate and take a close look at it. You will notice two important things. First, there are eight identical mounting holes around the perimeter of the frame (just as in plate D) and second, there are two long slots near the edges of the plate on one end (just as in plate D). Place plate E on top of our assembly and make sure the two long slots are directly over of the two long slots on our bottom plate. It should resemble Figure 2-19.

As you may have guessed, we need to thread more 5-mm mounting screws through the holes in plate E into our standoffs (see Figure 2-20). Start by manually placing the screws into the holes and then tighten them with a 2.5-mm Allen wrench (see Figure 2-21).

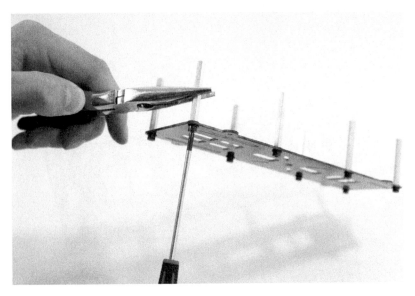

Figure 2-17. *Tightening the final 37-mm standoffs on plate D.*

Figure 2-18. *Plate D with all the 37-mm standoffs installed and ready for plate E.*

Figure 2-19. *Plate E on top of our assembly prior to adding the mounting screws.*

Figure 2-20. *Threading our first mounting screw through plate E into our 37-mm standoffs.*

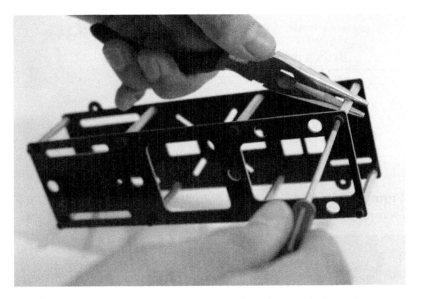

Figure 2-21. *Once you have all the screws threaded, make sure they are all nice and tight.*

 Lumenier QAV250

If you went with the QAV250 for your airframe, you will notice that the entire airframe is assembled in a similar fashion to our clean frame. The main difference is that the QAV has booms built into its bottom plate and only uses six standoffs to separate the two frame plates. This will create a lighter frame that is well suited for racing applications.

Good job! You have now built out the basis of both our subframes. If everything went as planned, your build should look like Figure 2-22. It's finally starting to look like a quadcopter, wouldn't you say?

Technically speaking, we need to stop our frame build at this point. There are electronics (covered in the next chapter) that need to be installed in both subframes, and that is much easier when they are not attached to each other. That being said, we are going to show you the final step of building out the frame now (it's a really simple one) and you can refer back to this chapter later in the book when we are ready for it.

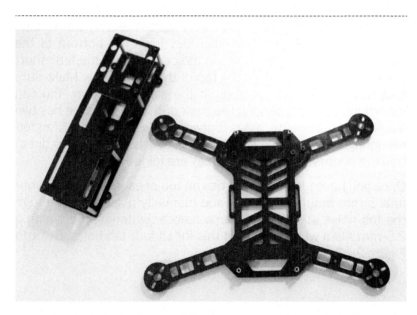

Figure 2-22. *Both of our subframes assembled and ready to be loaded up with flight components.*

Step 4: Attach the Two Subframes Together

Just to reiterate: this final frame-building step should not be performed if you are just getting to the end of Chapter 2. We need to install electronics in both of our subframes, and this step should be saved until we reach the end of Chapter 4 (autopilot install). Because this step is so simple, and for the sake of continuity, we will display it here and you can either refer back to this chapter or (more than likely) pull it from memory.

Begin by placing the clean frame on top of the dirty frame. Line up the four small tabs that stick out from the bottom of the clean frame (with 3-mm holes in the center) with the four short standoffs sticking up from the top of the dirty frame. Make sure that both frames are pointing in the same direction. You can determine this very simply: the back of the clean frame has two long slots along the sides between the back two standoff mounting screws, while the back of the dirty frame is the direction the booms will point when they are folded in.

Once you have placed the frames on top of each other, take your final 5-mm mounting screws and manually place them through the tab holes in the clean frame before tightening them with a 2.5-mm Allen wrench. Repeat this for all four tab holes. Refer to Figure 2-23 to see how it's done.

Figure 2-23. *Attaching our two subframes with a screw through the mounting tab (on the clean frame) into the short standoffs (on the dirty frame).*

That's it—you have attached the two subframes. That was easy, wasn't it?

What About the Camera Plate?

If you were paying close attention during this chapter (who are we kidding, of course you were!), you will have noticed that there is one additional plate that we have not used yet: plate F, otherwise known as the camera plate. This is going to be installed—along with the vibration isolation balls—later in the book. Cameras are discussed in Chapter 8, and we felt that was the best place to explain the details of what the camera plate does and how to install it.

 Watch It Being Built

You can find a detailed video (*http://gettingstarted withdrones.com/little-dipper-build/*) showing all of these steps as well as the results other readers have had by visiting the companion website.

3/Power Train

The *power train* is the group of components that generate power and transfer it in a fashion that creates movement. In your car, the engine creates power that is transferred to the transmission that sends it through the axle system and finally to the wheels which make the car move. Our quadcopter also has a power train system that helps us to create movement. Let's have a look at those components now.

Propellers

The best place to start is at the end, at the propeller. The prop, as it is usually called, is often compared to the tire in automobile power trains; just as the tire "grips the road" to create movement for the car, the propeller "grips" the air, moving the drone. And like tires for your car, props come in all shapes and sizes, and should be selected according to the rest of the power train that will drive them. There are three main specs you want to look for when selecting props: direction, size, and pitch.

Direction

Look carefully at the props in your kit. Notice that the propellers come in two different shapes with opposite angles of pitch. These props are meant to be spun in reverse directions from each other. Each type has the official name of *pusher* or *tractor*. Pusher props, sometimes called left-handed props, generate lift when spun in a counterclockwise direction. Tractor props, sometimes called right-handed, generate lift when spun in a clockwise direction. Together, these two props work to keep the aircraft balanced and level. We will discuss how this works in greater detail later in the book.

Pusher Props

Pusher props earned the name because they were originally designed for airplanes with rear-mounted motors (like the Wright brothers' first airplane) that would push the plane through the air, rather than pull it from the front. This is very similar to propulsion in a boat.

Size and Pitch

Propellers have a hub in the center that mounts directly to the motor, and two (sometimes more) blades that extend out from the hub. The size is pretty simple to define: it's measured in inches from tip to tip on a two-blade prop. For the Little Dipper, props in the 5- to 6-inch range are ideal, but some drones can use props up to 29 inches or more!

Pitch is a little more complicated to explain. Most people wrongly assume that it is a measurement of degrees that the prop is pointed up from a flat plane. In reality, the pitch is a measurement of how far the propeller would move forward if it were passing through a solid matter with one revolution. Think of the threads on a wood screw. A fine thread will move a shorter distance with one revolution compared to a coarse thread. Props work in the exact same way. A higher-pitched prop on the front of a plane, or even the back of a boat (in theory) would propel the craft further with one revolution than a lower-pitched prop. It's also harder to turn, just like that coarse screw!

It is outside of the scope of this book to discuss this in much more detail, but we will mention that the pitch of a propeller is rarely met in real life (see Figure 3-1). Something called prop slip always creates a difference between the distance the vehicle should have moved and how far it actually moved. If you are interested in that, a ton of websites have a wealth of knowledge on the topic.

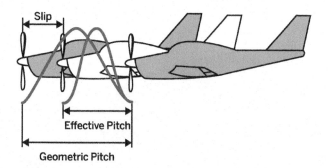

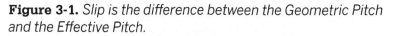

Figure 3-1. *Slip is the difference between the Geometric Pitch and the Effective Pitch.*

The higher the pitch, the more air will pass through the propeller with each revolution. Passing more air through the prop results in a higher level of thrust. This is great if you need more lift on your aircraft, but be warned that this results in less efficiency, because the aircraft's motors need more energy (in the form of battery power) to spin a higher-pitched prop. Moving up to a higher-pitched prop can reduce your overall flight time, as a result of gaining more lift.

Propellers come in all types of size and pitch combinations, each with their own pros and cons. Some are more efficient, while others generate more lift. Typically, the higher the pitch and/or the larger the prop, the more lift that is generated. If you have a heavy aircraft, or one that needs to fly at high speeds, a high-pitch or large prop would be ideal. However, if your needs are for a longer flight time with a lighter payload, something with a lower pitch will probably be best for you. Increased lift does come at a price, though: the efficiency of the prop.

The size and pitch specification is typically something like 0845P or 08X4.5P on the side of the prop itself, near the hub (see Figure 3-2). The first two numbers are the size, 8 (inches), followed by the pitch, 4.5°. The P shows that it is a pusher propeller. Some companies call pusher props left-handed props and may mark them with an L or R (for reverse) in the prop name instead of the P. Tractor props have no letter at the end; their designation would simply be 0845.

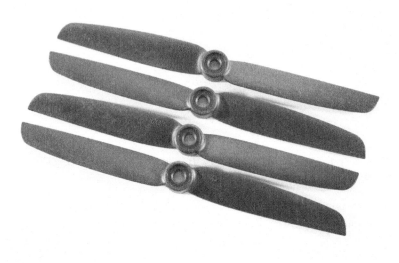

Figure 3-2. *Two pushers and two tractors.*

Efficiency Versus Maximum Thrust

The props we are using on the Little Dipper are 6 × 3. These have the advantage of providing more power from a larger 6-inch design, yet higher efficiency from the 3° pitch. If you are looking for more speed for your aircraft, try a higher-pitch prop like a 6 × 4.5. Remember: the higher the pitch, the more air is passed through the prop, at the cost of using more power from the battery.

Balancing Props

Here we find ourselves drawing another parallel between tires and propellers. Just as your tires need to be balanced to spin smoothly at high speeds, so too do your props. This is a delicate procedure, but one that is critical for anyone in this hobby to master. When you balance your props, you are making sure that each blade of the prop has an identical weight and will reduce the amount of vibration as it spins at high revolutions per minute (RPM).

The process is quite simple in theory: mount the prop you need balanced to a free-spinning axle. This free-spinning axle can be achieved in a number of different ways. Place a metal rod through the middle of your prop mounting hole and suspend that between two magnets, allowing it to spin from an almost frictionless fulcrum point. This lack of friction allows any heavy spot on the propeller to gently rotate toward the ground, under the effects of gravity.

Is the propeller perfectly motionless, or does it want to "lean" in one direction or the other? Using this technique to identify heavy spots in the prop, one can either apply additional weight to the light side, or subtract weight from the heavy side. Keep in mind that these are typically very small adjustments. Adding weight might come in the form of a few small clips of Scotch tape or label stickers, while subtracting weight is usually done with sandpaper and a bit of patience. When fully balanced, the propeller should be able to be moved to any position without one blade or the other leaning toward the ground.

If the thought of building a prop balancer sounds intimidating, don't worry. Several companies make balancers of all shapes and sizes. The model that we use, shown in Figure 3-3, is one that we have had as long as we have been working with RC aircraft.

Modern-day propellers come pretty close to perfectly balanced from the factory, but it's always a good idea to double-check them yourself. Usually the better quality the prop is, the more likely it is to come balanced from the factory. Cheaper props may save you a little money, but you will have to spend the time to balance them yourself.

Figure 3-3. *When perfectly balanced, the prop should be able to sit in the balancer in any position without falling in one direction or the other.*

Balancing Video

Still not sure how to go about balancing your props? We have a video (*http://bit.ly/prop_balancing*) that shows you exactly how it's done, step by step.

Motors

One major difference between our quadcopter's power train and your car's is that we use a direct-drive system rather than a transmission. With your car, you have one main source of power: the engine, which is connected to the transmission, which divides that power among the wheels that need it. With direct drive, each wheel, or prop in this case, is directly attached to its own power source. For our quads, the power source is a brushless motor.

Sizes

Most modern brushless motors in the hobby industry are identified by the width and height of the motor housing (see
Figure 3-4). A 2216 motor is 22-mm wide and 16-mm tall. Of
course, not every company subscribes to the exact same nam-
ing conventions, so it's always a good idea to refer to a spec
sheet when possible.

Figure 3-4. *These 2204 motors are rated at 2300 kV and work
with 5- or 6-inch props.*

kV Rating

Another spec that you have probably seen stamped on the side
of brushless motors is the kV rating. This tells us the RPM of the
motor for each volt of electricity fed into it. This measurement is
under zero load on the motor, so actually RPM would vary
depending on friction and load. You can now see how this rating
is a sliding scale for RPM depending on the battery voltage. A
higher voltage battery will make the motor spin faster, but the
kV rating is still the same.

Calculating RPM

A 900 kV motor with a battery putting out 12 volts will spin at 10,800 RPM (12 x 900) under no load. The same motor using a larger battery (16.8 volts) would spin at 15,120 RPM.

Pairing with the Right Props

Each motor will perform at different levels depending on the propeller it is paired with. These days, the guesswork is usually minimal, because most of the motor manufacturers post the recommended prop specs for their motors. To determine the correct prop for your build, you should have an idea of what your drone's total weight will be once you are done. The heavier you plan your build to be, the larger prop you should use based on the manufacturer's suggestions.

Total Lift

A common term you will hear in relation to motor and prop combos is their total *lift*, which is the amount of upward thrust a motor/prop combo generates in the real world. This is typically documented with a weight value that represents the lift generated at 100% throttle command. (Most manufacturers provide this information.) The 2204 motors we used on the Little Dipper build can generate a total lift of 539 grams with a 6 × 3 prop. If we changed props, the lift number would change. Obviously, this is measured for each motor and propeller. To calculate the lift of the entire copter, multiple your results by the numbers of arms on your copter (four for our quad). For the Little Dipper, that works out to be (4 × 539 grams) 2156 grams of lift!

Calculate Your Payload Capacity

One spec that every UAV designer has to define is the payload capacity. This is the difference between the aircraft's *total lift* and its *all up weight* (AUW). The AUW is exactly what it sounds like: the weight of the aircraft as it is outfitted for flight, with motors, battery pack, computer, and so on. Now that we understand how to calculate our total lift, we can weigh all of our components to find our AUW and determine our aircraft's payload capacity.

Electronic Speed Controllers

Electronic speed controllers (ESCs) are small electronic circuits that are used to independently control the speed and direction of each motor on our quad. Four ESCs are installed on our aircraft, each designed specifically for use with brushless motors. They work by converting power from the main flight battery into a sequence of electrical signals that are sent across three different wires to the brushless motor. That sequence controls the speed, rotation, and even braking ability of the motor. The required speed for each motor is communicated to each ESC from the flight controller. We will have more on that in the next section.

Classification: Amps and volts

There are generally two main specs you need to look for when buying ESCs: amperage and voltage. The number of volts an ESC is rated for determines what size battery you can use with the ESC (we will discuss batteries in more depth in "Flight Battery" on page 50). For now, we will just point out that the unit we will be using in our example is a three-cell battery rated at 12.6 volts. Remember the motor spec sheets we mentioned while discussing total lift? Those same specs generally contain the current draw in amps as well. That information is what you use to determine what size ESCs you need for your desired motor and props.

SimonK firmware

The types of ESCs used in hobby-grade drones were originally designed to be used in RC airplanes. In an effort to upgrade performance, RC enthusiast Simon Kirby developed an open source firmware for hobby-grade ESCs, appropriately called the SimonK firmware. His upgrade provided a faster response time for the motors, which greatly improved the aircraft's stability. Most major ESC manufacturers now offer SimonK ESCs specifically tuned for multirotor use (see Figure 3-5).

Figure 3-5. *The ESCs we use in our example build are rated for 12 amps and up to 16.8 volts (4S battery).*

Flight Battery

One of the most important technological developments that helped give rise to the current civilian drone is the lithium-polymer (Li-Po) battery. Very similar to those used in smartphones, Li-Po batteries have a much greater capacity-to-weight ratio than older-generation nickel-cadmium (NiCD) and nickel-metal hydride (NiMH) batteries. That weight savings was key to helping UAVs get off the ground.

Capacity and voltage

Most modern Li-Po batteries are classified according to their capacity and voltage. All Li-Po batteries achieve their end voltage by wiring up a series of smaller battery cells inside the main battery unit. Each cell is rated for 3.7 volts (4.2 fully charged). That means a three-cell battery, such as the one in our demo, is rated for 11.1 volts (3 x 3.7v) and can reach 12.6 volts (3 x 4.2v) when it is fully charged.

The unit of measurement for capacity is the *amp-hour* (Ah). This describes how long the battery charge will last under certain loads: a 10 Ah battery powering a 1-amp device should last around 10 hours. The same battery with a 5-amp load will last around 2 hours. All of our batteries will come with a milli-amp hour (mAh) rating that determines the capacity of our battery (see Figure 3-6). A battery rated at 2200 mAh has a higher capacity than one with only 1500 mAh. In theory, this battery will last more than 30% longer than the 1500 mAh. In the real world of aviation, though, that might not be the case. A 2200-mAh battery weighs more than a 1500-mAh battery, which means it will have to expend more power to lift its quadcopter. We will discuss this delicate balance more throughout the book.

Although Li-Po batteries helped propel our hobby forward, they still had some drawbacks. Li-Po batteries become unstable under certain circumstances and can catch fire!

For this reason, you should always use great care when handling Li-Pos. Replace damaged batteries, and never use a battery that has been punctured in any way; that is a recipe for disaster! Use fireproof bags when charging Li-Po batteries. Simply hook your battery up to the charger as you normally would, then place the battery inside of the fireproof bag and close the opening around it as much as possible. It is always a good idea to keep an eye on your batteries while charging.

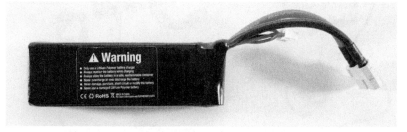

Figure 3-6. *Three-cell 2200-mAh main flight battery.*

Leave a little in the tank

The rule of thumb is to never discharge your battery below 3.2 volts per cell. That means that for a three-cell battery, discharging below 9.6 volts will damage it. We find that setting an alarm for 3.5 or 3.4 volts per cell lets us land with just enough in the tank (about 20%) to keep our batteries happy. Also, avoid storing batteries that have a full charge. Most decent chargers have a storage setting you can use that brings it down to a nice 3.8 volts per cell. This will greatly increase the life of your batteries.

Step-by-Step Build Instructions

For this portion of the build, you will need (see Figure 3-7):

- Soldering iron and solder
- Helping Hands or some other clamping system
- Heat gun
- 12 pairs (male and female) of 2-mm bullet connectors
- Several inches of 1/8-inch heat shrink
- Wire cutters/strippers
- Allen wrenches
- Small zip ties
- Double-sided foam tape
- Scissors

Figure 3-7. *The materials list for this chapter is pretty serious!*

Step 1: Mount the Power Distribution Board (PDB)

The very first thing we need to do is secure a location in our dirty frame for the PDB. At the time we did our build, we made our own PDB from copper-clad G-10 (see Figures 3-8 and 3-9), but there are many small inexpensive versions on the market that fit the need. This one fits perfectly in the middle of our dirty frame. Start by applying a couple of small strips of double-sided tape to the back side of the PDB (see Figure 3-10).

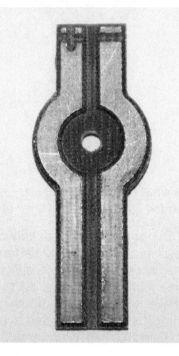

Figure 3-8. *Our DIY power distribution board cut from copper-clad G10.*

Figure 3-9. *This is what our PDB looked like—we tinned a few solder spots and added liquid tape for a layer of insulation.*

Now press the PDB into place in the middle of the dirty frame where it will be easily accessible by the battery lead, ESCs, and any extras that need access to power (see Figure 3-11). There is a 3-mm hole in the middle if you would like to also add a screw for extra support. We found that the double-sided tape did a very good job holding it in place and opted to not use one. If you do use a screw, try using a small nylon screw and nut, which will save on weight and also does not act as a conductor.

Figure 3-10. *Applying double-sided tape to the back side of the PDB.*

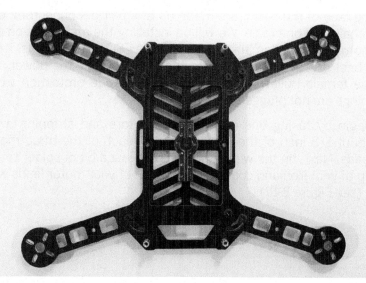

Figure 3-11. *The PDB is pressed firmly into place.*

Step 2: Solder on the Bullet Connectors

Th is step is somewhat optional, but it can make the install a lot easier. Bullet connectors allow you to plug and unplug the ESCs and motors into each other rather than soldering them directly. The pros to using them include ease of use during maintenance, troubleshooting, and upgrades. The cons include failure due to loss of contact. If a bullet connector fails, it can cause a crash on a quad (one motor out of four stops spinning and you fall like a rock). With this list of pros and cons, you can understand why people have sharp opinions about these connectors in both directions. We will let you decide for yourself if you want to use them, but this book will assume that the user has them installed. If you decide not to use them, we recommend that you directly solder your connections and seal them with heat shrink. Just make sure your connections are right before you fire up that iron!

Pull out the third-hand helper and turn on your soldering iron—it's time to get to work!

Bullet connectors, like almost every other type of connector on the face of the earth, are made up of a pair of connectors: one female and one male. We will be installing the male ends on our motors and the female versions on the ESCs. This is considered a best practice, as the ESC is the end providing the power and the female bullet will be shielded to provide protection when things are not plugged in.

Begin by taking one of your four motors and stripping away about 1/8 inch of the insulation from each of the three motor leads. Next, tin the wire tips by adding just a bit of solder to the tip of your iron and coating the outside of your motor leads with it (see Figure 3-12).

Figure 3-12. *The motor leads are stripped, tinned, and ready to take the male bullet connectors.*

New to Soldering?

You are reading through a *Make:* book, so we assume that you have used a soldering iron at some point in your past. Maybe it was only once, maybe every day, but we do assume a basic understanding of how it works. If you are entirely new to the concept, or just want a refresher course, check out our in-depth soldering tutorial (*http://bit.ly/soldering_tutorial*) on the website.

Next, we are going to get out our third hand and use it to add the male bullet connectors to our motor leads. Clamp one bullet connector into one of the alligator clips with one of the tinned motor leads in the other. Take your time and make sure that you can position both of these parts in a comfortable way so that you can easily access them with your soldering iron. Once you have everything configured as needed, place your iron on the outside of the bullet connector, allowing it to heat up for just a few seconds before applying some solder to the inside of the connector where the wire sits. Refer to Figure 3-13 for more info.

Figure 3-13. *Everything is in position and we are ready to solder our first connector on.*

Apply a moderate amount of solder without going overboard. Once the end of the bullet connector appears to be close to full, remove the iron and solder. Allow the connection to cool for a few moments before removing it from the third hand (see Figure 3-14).

Congrats: you just finished your first solder job on this build! Does it feel good? We hope so, because you have a ton more to do. Let's get to it!

Once the solder has cooled, remove the motor lead and bullet from the third hand and repeat those steps to solder the other two motor leads and bullet connectors. After your first motor is done, repeat the steps for the remaining 3 motors. When you are done, you should have 4 motors with 12 male bullet connectors soldered to all of their motor leads (1 on each lead).

Figure 3-14. *Our solder is cooling while being held in place with the third hand.*

Now it's time to insulate our connection. For that step, we will need our 1/8-inch heat-shrink tubing and a heat gun (or hair dryer if you don't have one). Cut three 1/2-inch sections of heat shrink and loosely fit them over your newly soldered bullet connections. You want to place the heat shrink across the back raised portion of the connector, as the front male plug part will be inserted into the female plug (see Figure 3-15). If we have heat shrink blocking the connection, it can lead to an unreliable plug. Feel free to plug the male and female plugs into each other a few times and take note of where the connections take place. This will help you better understand where you should—and, more important, should not—have heat shrink placed over your connection.

Figure 3-15. *Heat shrink is in place and ready for the heat gun.*

Once you have the heat shrink in position, get your heat gun ready. We will gently apply some heat to the heat shrink (see Figure 3-16). Try to position the wires so that you can heat one at a time, especially if you are new to this process. If you are using a hair dryer or heat gun with a lot of wind displacement, be careful that the moving air does not move your carefully positioned heat shrink. Apply this process to all three leads and you should end up with something that looks like Figure 3-17. It only takes a few seconds to shrink the tubing. Once it's tight around the connector and wire you are good to move on.

Figure 3-16. *Using our heat gun to shrink our tubing for insulation purposes.*

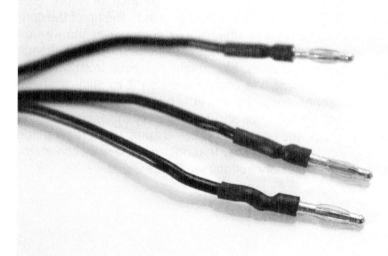

Figure 3-17. *The finished product.*

We are starting to make some real progress. At this point, you have four motors with bullet connectors firmly soldered onto each of their leads, which are carefully insulated with heat shrink. Now it's time to do it all over again with the female connectors on the ESCs. Depending on what ESCs you have bought, there may already be bullet connectors soldered on. If yours do have connectors in place already, check that they work with your male motor bullets. If everything seems to connect nice and snug, skip the rest of this step.

As with your motors, find the three black leads coming out of your ESCs (not the servo plug, the raw wires), strip about 1/8 inch of the insulation, and prepare the wire for soldering. We are going to follow the exact same steps that we did for the motors with the only difference being that we are soldering the female connectors on this time (see Figure 3-18). Walk back through the previous steps if they're not already ingrained in your head from having done it so many times!

After you have all the connectors soldered up and are ready to begin insulating them, take note of the difference in area that needs to be insulated between the male and female plugs. Rather than the 1/2 inch of heat shrink that you used for the male plugs, the females will require an inch or more (see Figure 3-19). The heat shrink should go just to the tip of the connector without going over while still extending over the wire on the other end. Keep in mind that your heat shrink will change shape a little as it shrinks, so it might pull back from the edge when you apply heat. We usually position it to stick past the edge of the connector just a tiny little bit in anticipation of it lining up perfectly after it has shrunk. If you try this technique and it sticks out over the edge after it has shrunk, use a razor blade to carefully cut any parts away that obstruct the male connector from making a solid connection (see Figure 3-20).

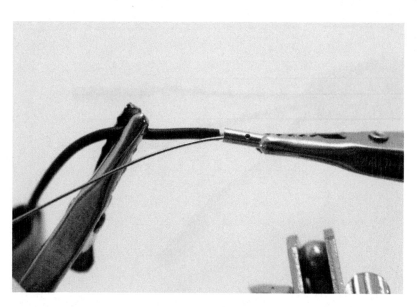

Figure 3-18. *Soldering up our first female bullet connector.*

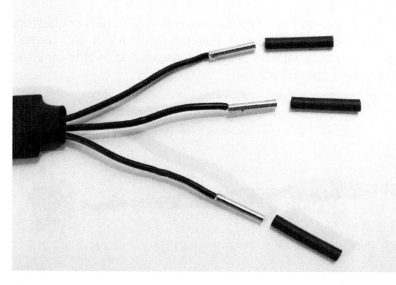

Figure 3-19. *Be sure to cut the heat shrink to the right length for the female connectors.*

Figure 3-20. *After applying a little heat, everything looks as snug as a bug in a rug!*

At this point, we have four motors with male bullet connectors and four ESCs with female bullet connectors. If you haven't done so already, let's try plugging them into each other and see how they fit (see Figure 3-21).

Figure 3-21. *Success: everything fits like a glove.*

Fantastic—you are well on your way to getting your power train installed on your drone. At this point, put down what you are doing and grab your favorite refreshment from the kitchen. You deserve it!

Step 3: Mount the Speed Controllers

Electronic speed controllers are typically mounted in one of two ways: either on the frame itself, or out on the booms near the spinning propellers in order to get additional cooling from the downdraft of the props. Because our booms fold on this particular frame, we will be mounting the speed controllers on the inside of the dirty frame. Use double-sided tape to mount the speed controllers. Apply a small strip about 1/2-inch wide to a single side of the ESC (see Figures 3-22 and 3-23). Check to see if one side is more flat than the other. Sometimes ESCs can have a large round capacitor that sticks up on one side. If you find that to be the case with your ESC, apply the tape to the other side and stick the capacitor up so that the tape can have the most amount of surface coverage as possible.

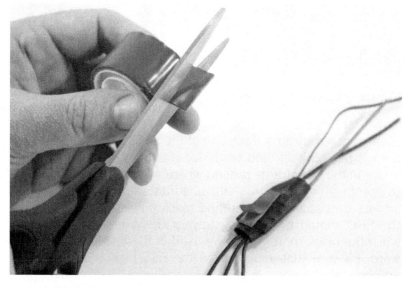

Figure 3-22. *Apply a small strip of double-sided tape of the flattest side of the ESC.*

Figure 3-23. *Repeat for all four ESCs.*

Now that we have the tape applied to the ESCs, let's position them in our subframe. Take one of the ESCs and make sure the tape is peeled back and ready for mounting. Locate the open space in the dirty frame around where we installed our PDB. Try to mentally separate this into quadrants and place each ESC into its own space. The red and black power leads coming off the ESC should be pointing toward the center of the frame, while the black motor leads we used in the last step point outward. Be sure that you position the ESC high enough so that another can fit underneath it. Refer to Figures 3-24 and 3-25 for more info.

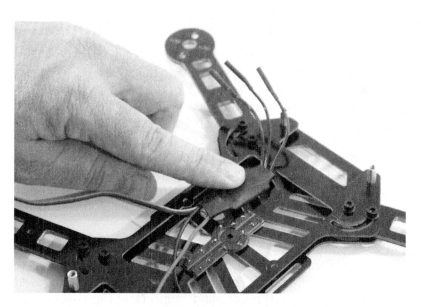

Figure 3-24. *Position the ESC and then press it firmly into place, making sure there is enough space below it for another unit.*

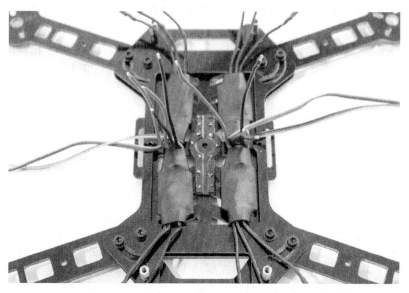

Figure 3-25. *When all of the ESCs are in place, it should look like this.*

Step 4: Solder up the Power Supply

The next step is to solder up the power supply. The overall concept here is to connect the positive and negative leads (red and black wires, respectively) from each of the ESCs in a parallel circuit. If you aren't familiar with a parallel circuit, that's OK. It's a pretty simple concept. It simply means that all of the red wires (positive) are joined together in one connection while all of the black wires (ground) are on another connection. If you look at Figure 3-9, it becomes very clear how that works. We have one strip on the board for positive leads and one for ground. All of the ESCs as well as our main battery lead will connect to the PDB.

Let's start with a single ESC as the first example. Take the red wire coming out of your ESC and determine how long it needs to be in order to effectively reach a positive circuit on the power supply (in our case, the left-hand strip). Now clip that wire to that length (or just a tiny bit longer, just in case) and strip off 1/8 inch of insulation from the tip. Now tin the exposed wire with your soldering iron and get it ready to be attached to the PDB (see Figure 3-26).

Once you have the positive lead tinned, make sure the tip of your soldering iron is nice and clean before loading it up with a little more solder. Next, take your needle-nose pliers and use them to hold your ESC lead onto the PDB at the point where you want to make the connection. Make sure you are on the correct PDB circuit. This is our positive lead, so make sure it's on the positive circuit. Finally, apply your hot iron to the top of the positive lead, sandwiching it between your iron tip and the PDB. If you have applied enough solder to all the components, they should all melt together with no problem. Once that happens, remove your iron while continuing to hold the lead for a few more seconds with the pliers. If you pay attention to the solder, you will see it cool in a matter of seconds. It will take on more of a matte finish look and less of a liquid appearance. Once this has happened, you can remove the pliers and check the connection. If it appears to be loose at all, repeat the necessary steps until you have a solid solder joint (see Figure 3-27).

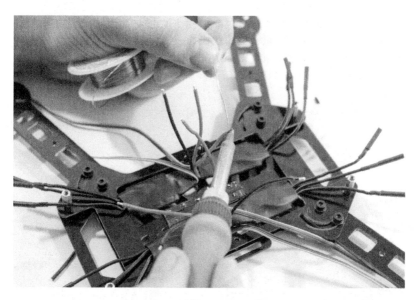

Figure 3-26. *Tinning our first ESC power lead before attaching it to the PDB.*

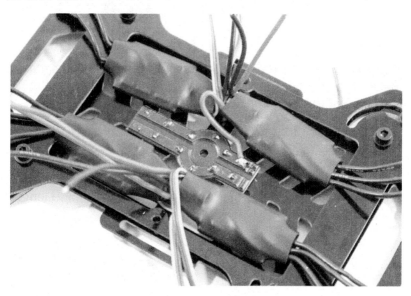

Figure 3-27. *Our first lead is soldered and seems to be a solid connection.*

After you have the first positive lead in place, repeat the same steps for your negative lead on the same ESC. The only thing you should do differently is connect the lead to the negative circuit on the PDB (in our case, the right-hand strip); see Figures 3-28 and 3-29.

Figure 3-28. *Solder the negative lead in the same fashion, making sure you connect it to the negative circuit seen here on the strip closest to the bottom of the image.*

Awesome—your first ESC is connected to the PDB! Now do the rest in the same manner (see Figure 3-30). Take your time and think about where you will position the wires for all of the ESCs. Don't make the mistake of cutting one of your wires too short in an attempt to save space. It's better to leave a little extra length at first.

Figure 3-29. *Our first ESC is all ready to get power from the PDB.*

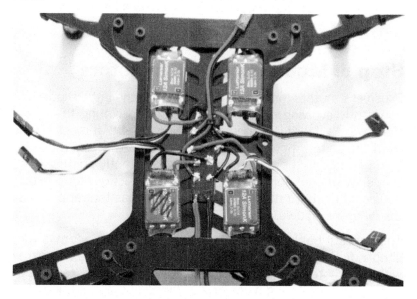

Figure 3-30. *Example of another build from our fleet with all the ESCs connected to the PDB.*

We are almost done with our soldering iron. Is it getting hot in your workshop yet? The only thing left to solder is the main battery lead. This is attached in the exact same fashion as the ESC leads. Clip some insulation from the end, tin the wire, hold it in place with the pliers, and apply some heat. Be sure that you are connecting to the right circuits and that your solder joint is nice and solid.

Future-Proofing your Drone

Although you are only adding a certain number of components to your aircraft right now, who knows what the future will hold? It's good to plan ahead now by adding an extra unused power lead to the PDB for items you want to include down the road. In Figure 3-30, you will notice that we have done exactly that. Try adding a JST power lead (readily available online for a few cents) to your power circuit and leaving it tucked neatly between the clean and dirty frames. Then when you want to add something like a video transmitter, all you have to do is pull that plug out and tap into your power right there. No need to get the soldering iron out again!

Step 5: Mount Brushless Motors

Brushless motors for small drones such as the Little Dipper are constantly evolving. At the time we started writing this book, there were a handful of models on the market that most people used. Just a few months later, there were dozens, and more are being added all the time. Due to this constantly shifting landscape, we are not going to give specific instructions for one particular model but rather the overarching ideas that apply to all the different models.

The concept behind mounting your motors to the boom is very simple; find the appropriate-length mounting screws that came with your motor (there could be multiple lengths) and feed them through the mounting holes on the round end of the boom and into the threaded holes on the bottom of your motor. Once you have them hand-tight, work each opposite screw turn for turn the same way that you would on a car wheel while changing a flat.

Pay Attention to the Thread Direction!

Notice that your motors have a shaft that holds the prop in place. This shaft is threaded, and a nut of some type (called the *prop nut*) fits over that thread and puts pressure on the prop. In the early days of small drones, those threads were all standard clockwise threads. Because our motors spin in both clockwise and counterclockwise directions, manufacturers realized that if they created motors with standard and reverse threads, they could use the spinning prop direction to help keep it tight. Make sure that you always have a thread that screws on in the opposite direction from that in which the prop will be spinning.

Clockwise spinning motors should have a reverse-threaded shaft. Counterclockwise motors work best with a standard thread.

One catch: not all manufacturers make their motors this way. Many still only use a standard thread on all of their motors. This will still work fine; just be sure to tighten it well and check it regularly (as you should anyway).

The moral to the story is to always know what type of threads your motor shaft has and be sure you are installing them in the right location. If you need to confirm the direction that your motor will be spinning, do so before attaching it to the frame.

For our Little Dipper build, our motors will spin clockwise on the NE and SW motors while the NW and SE motors spin counterclockwise.

Start your mounting process by laying the motor flat on the top of the boom while lining up the mounting holes in the bottom of the boom with the threaded holes in the bottom of your motor (see Figure 3-31). Make sure that you have a motor with the correct thread direction for the anticipated motor direction. The motor leads should run down the length of the boom. Make sure they do not run in any other direction.

Figure 3-31. *Our motor is sitting squarely on the boom ready to be attached.*

Now manually feed the first screw through the boom's motor mounting hole and into the threaded holes on the bottom of the motor. Once the first one is done, feed the opposite screw in and keep going until you have all four in place (see Figure 3-32).

Once you have all of the screws manually fed into place, tighten them turn for turn on opposing screws, similar to lug nuts on a car (see Figure 3-33).

Now that you have the first motor mounted in place, move on to the remaining three. Don't forget to keep an eye on what direction threads you are installing!

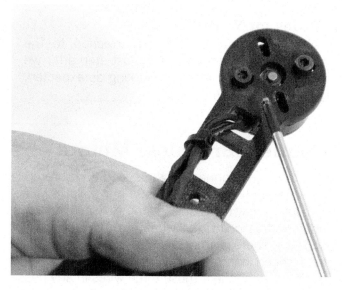

Figure 3-32. *Feed the screws in the same way you would a set of lug nuts on a car to ensure an even tightness all the way around.*

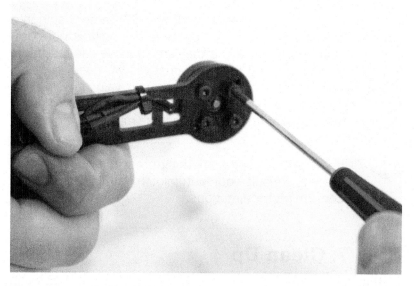

Figure 3-33. *Tighten the screws turn for turn to ensure that they are all even.*

No Props for You!
Be sure the propellers are not attached to the motors yet. That will be our very last step after we have confirmed everything is working as expected. This is an important safety step.

Step 6: Connect Brushless Motors

Our brushless motors will connect to the speed controllers via the three black wires that that we attached our bullet connectors to earlier. If you're new to brushless motors, you may notice something funny at this point: the wires are not labeled. But there's a reason why. There is no wrong way to connect a brushless motor to a speed controller, only different directions of motor rotation. You can connect those three wires in any possible combination and it would never be "wrong" in the sense that you are going to damage the motor; it will simply spin in one direction or the other.

Our ultimate goal is to make the NE and SW motors spin in a counterclockwise direction. But because our build is not complete yet, just hook them all up the same and we can test them later in the book to find out what changes we need to make. Now it becomes pretty apparent why there is value in using bullet connectors! If you had to do this step with soldering directly between the components, it would become much more difficult!

If you are building something other than our example kit, consult your autopilot manual in order to confirm motor directions.

Step 7: Clean Up

At this point, it's a great idea to use a couple of cable ties and clean up your wiring job. It's also helpful to place identifying tabs on the servo leads coming out of the speed controllers before you button everything up. This will make your life much easier when working on subsequent projects later in the book. We usually use a fine-tip marker or paint pen and label the motor number on the ESC lead.

Wrapping Up

At this point, you should have your speed controllers and motors permanently mounted. Make sure everything is firmly mounted with little risk of coming loose. Also check all soldering in your power harness/power distribution board to make sure nothing is loose. The connection between your motors and speed controllers should be temporary at this point. We will return to the motor wiring later in the book when all our components are installed.

4/Flight Controller

What Is a Flight Controller?

Basically, the flight controller is the brain of your drone. It measures the performance of your aircraft through an array of various sensors, hundreds of times per second, and then manages all the microadjustments that are needed on each motor to keep your aircraft stable in the air. Think of what happens when you walk down the street. You are not thinking about the variations needed in your gait due to an uneven sidewalk or headwind. You just walk, and your brain comes up with those details and executes them while your only thought is "forward." The flight controller acts in the same manner. When you give your drone the command to move forward at 10 miles per hour, the flight controller is taking your overall command and breaking it down into hundreds of commands per second for each of the motors in order to achieve said command. If the drone encounters some type of resistance, such as a headwind, the flight controller will attempt to make up the difference needed to still achieve your command without extra input from the pilot.

Open Source Versus Closed Source

You can break all the flight controllers in the world down into two categories: open source and closed source. Some popular open source flight controllers include APM (AudruPilot), Open Pilot's CC3D, Sparky, and various flavors of MultiWii. These projects all have one thing in common: anyone can download the build files and software needed to make their own. We highly encourage you to spend a little time online to check out all these projects. They all have great communities built around them and offer a vast wealth of knowledge on UAVs.

Closed source flight controllers come in all types of configurations. Some of the more popular models are the Wookong and Naza by DJI as well as the Super-X and Mini-X by X-Aircraft.

These systems are usually proprietary and do not offer the user the chance to modify the software code itself.

Sensors

The heart of the flight controller is the *inertial measurement unit* (IMU); see Figure 4-1. This unit contains a suite of sensors to help the flight controller monitor the activity of the aircraft. The most typical sensors include an accelerometer, gyroscope, and barometer. These measure the acceleration and rate of rotation of the aircraft, and the air pressure (which can often be converted to the aircraft's altitude). These three sensors provide all the data that the autopilot needs to keep the aircraft stable in the air.

Additional sensors are also used outside of the IMU for advanced functionality. These include GPS receivers, magnetometers, optical flow sensors, and air speed sensors. As time moves forward, we will see even more sensors introduced into flight controllers to help with tasks such as Sense and Avoid, computer vision, and artificial intelligence (AI).

Figure 4-1. *This flight controller has a built-in IMU and ports for external sensors.*

Here is an overview of the APM and all of the inputs/outputs (I/Os):

Inputs 1–8
> These ports will be connected to our radio receiver. This collects our flight instructions from the ground and delivers them to the APM.

GPS
> This port will be connected to our GPS sensor.

I2C (inter-integrated circuit, pronounced I-squared-C)
> This is a standardized serial computer bus that is used for many types of peripherals. If you have used an Arduino before, you may be familiar with the I2C protocol. We will be using this port for our external compass.

PM
> This is our power module port. This allows us to monitor our power supply.

JP1
> This is a jumper port used to tell the APM if there should be a power module in use.

Outputs1–8
> These will supply the signal we need for our electronic speed controllers. Alternatively, these could send a signal to a set of servos if you used APM for a model airplane or inputs to a gimbal controller if you asked APM to control a camera mount on your drone.

Telem
> Our telemetry radio plugs in here.

A0–A11
> Just as in an Arduino, these are our user assignable analog pins. These will not be used in this book.

Flight Characteristics

While they are all adjustable to a certain extent, each flight controller usually brings its own set of flight characteristics to the table. Some controllers are more soft and mushy and allow the copter to just loft about in a very fluid manner. Those types of flight characteristics are more sought after by aerial photographers or videographers with larger aircraft. Others may tend to be more snappy and have a greater feeling of authority, such as those used by first-person view (FPV) racers.

Software Assistant

All modern flight controllers have some type of software interface (see Figure 4-2). This allows the user to gain access to many important parameters within the flight controller. While some software assistants are more robust than others, they all generally control the aircraft configuration, communications setup, and autolevel gains. Software assistants can connect through Bluetooth, USB, or telemetry radio, and are developed for both mobile and desktop platforms.

The assistant that we will be using, called Mission Planner, also includes a built-in ground station. This allows the application to display important information that is sent back from the aircraft, such as geolocation and sensor readings. Access to this information on the ground can be a tremendous help for certain types of flights—such as mapping or agriculture monitoring missions—and really helps during any type of troubleshooting process.

If you are following along in the book with the companion kit, download Ardupilot's Mission Planner software application (*http://ardupilot.com/downloads/*). If you are on a Mac, try using APM Planner instead.

Figure 4-2. *The Tower flight controller app from 3D Robotics running on an Android tablet.*

Step-by-Step Build Instructions

For this portion of the build, you will need:

- One inch of foam double-sided tape
- Scissors
- Fine-point paint pen or Sharpie
- Small zip ties

Step 1: Mount the Flight Controller

The APM flight controller will be mounted on the bottom plate of the clean frame. Use two small pieces of double-sided tape at either end of the flight controller to attach it firmly to the frame (see Figures 4-3 and 4-4). Avoid using one large piece of tape, which can be difficult to remove later. The tape does have some vibration isolation qualities, so generally the thicker tape you can find, the better (up to a degree; there's no need for 1/2-inch thick tape). Make sure that the forward arrow on the APM flight controller is pointing forward on the UAV, and that the flight controller is positioned directly in the center of the frame before pushing it into place (see Figure 4-5).

Which Way Is Forward?

Recall from previous chapters that our airframe has a defined front and back. The back has two long, narrow battery-strap slots between the two rear standoff mounting holes. These will appear on both the top and bottom plates of the clean frame. Another indicator is that the booms fold *backward* on the frame. Yet another indicator is that the large hole that the camera mount will cover (covered in Chapter 8) in the clean frame sits in the front. Make sure that your autopilot is pointing toward the front—look for the forward arrow on the APM case—and is as centered as possible. Pointing it in the wrong direction will make for a very short maiden flight!

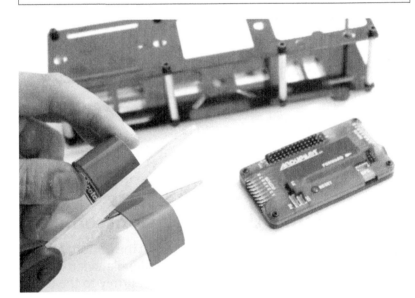

Figure 4-3. *Cut off about an inch of double-sided tape to use in the mounting of the APM.*

Figure 4-4. Cut your tape in half and place each portion at one end of the APM, then peel off the backing.

Figure 4-5. With the adhesive tape exposed, position the APM in the center—making sure it is pointing in the right direction—and firmly press into place.

Step 2: Wire up the Flight Controller Outputs

As we have discussed, the flight controller is nothing more than a microprocessor that takes an array of inputs, runs calculations on those inputs, and then sends out the appropriate signals to keep you airborne. At this point, the only things we have to wire up to the flight controller are our motors and ESCs (outputs). Take a look at the outputs labeled 1–8 on the APM. Notice that each output has three header pins that are vertically stacked under the number that labels it. Those three pins are for the signal, power, and ground found in each servo lead on your ESCs. They are even labeled on the right of the pins (S, +, -); look directly at them from the back of the unit.

Wire Color Standards

The three-wire standard that you see on the your ESC is sometimes called a *servo wire*. This always contains a power, signal, and ground wire with the female header connector that you see. What can change is the color of the wires. As a rule of thumb, the power will *always* be red, but the other colors can fall under one of two styles. They will either be black = ground/white = signal, or brown = ground/orange = signal. Learning this now will save you headaches down the road.

Outputs 1 and 2 are for the NE and SW motors, respectively. Identify the speed controller that handles each motor, and plug the servo lead from that ESC into the appropriate output port. Make sure that all ground (black) wires are at the bottom of the servo lead when plugged in. Repeat this step, adding the NW motor to output 3, and SE motor to output 4. This is clearly illustrated in Figure 4-6.

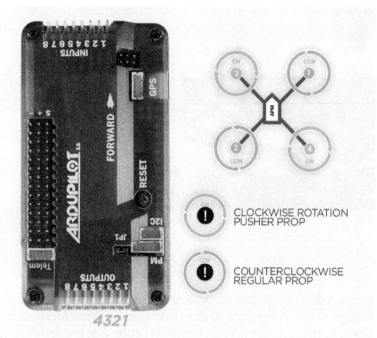

CLOCKWISE ROTATION
PUSHER PROP

COUNTERCLOCKWISE
REGULAR PROP

Figure 4-6. *Each motor has a dedicated output and rotation direction; the APM unit and the APM diagram are both pointing in the same forward direction.*

To properly route these cables, take a small paint pen or Sharpie and label each ESC plug with its corresponding output number. Once they are clearly labeled, align the tips of all the plugs and place a small zip tie or two along the cable length just below the plugs. This will help keep everything neat and organized. Now push that group of cables through one of the four rectangular holes in the back of the clean frame (near the APM output pins). See Figure 4-7 for a detailed photo of this process.

Now, simply plug each of the ESC plugs in to the APM output that you labeled earlier. If you followed these steps correctly, you should have a nice neat connection between your APM and the aircraft's ESCs, similar to Figure 4-8. Notice the attention we paid to keep the wiring as neat as possible. A clean build is easier to work on.

Figure 4-7. *Our ESC leads route nicely from the dirty frame up through a hole in the back of the clean frame, giving them easy access to the output ports on the APM (note that the top clean frame plate was removed so we could get a better photo—it is not required to do this).*

 If you are using a flight controller other than the APM, check your instruction manual for the proper motor output sequence. APMs always start with output 1 and count up depending on how many motors your drone uses. A hexacopter always uses outputs 1–6, while an Octocopter always uses outputs 1–8.

Figure 4-8. *All of our ESCs are plugged in to the APM outputs (Note: the top clean frame plate was removed so we could get a better photo. It is not required to do this).*

Step 3: Attach the Two Subframes to Each Other

Now that we have our ESCs wired into the APM, we can go ahead and attach the two subframes to each other. If you remember, we covered this step at the end of Chapter 2 while we were building out the frame. If you don't recall this step, refer back to "Step 4: Attach the Two Subframes Together" on page 38. As a quick review, you are running four 5-mm mounting screws through the mounting tabs on the clean frame into the short standoffs located on the top of the dirty frame. Figure 4-9 may jog your memory.

Figure 4-9. *Finally, we are ready to attach our two subframes to each other!*

Wrapping Up

By now we have our flight controller firmly mounted to the clean frame, and have outputs going to the four motors. Double-check all connections and make sure that the appropriate wires are going into the appropriate ports. Double-checking now can save you headaches later.

5/GPS, Compass, and Battery Monitor

In this chapter, we are going to cover three accessories that we will add to our drone. Each of these components on its own would make a pretty small chapter, so we thought we would cover all three in a single chapter.

GPS

One feature clearly draws the line between a drone and a model aircraft: the ability to operate using GPS (see Figure 5-1). The addition of this satellite navigation technology allows for a level of control that was simply not possible before. This extra control allowed drone designers to develop flight modes for specific types of flights.

Figure 5-1. *3DR uBlox GPS unit with built-in compass.*

Flight Modes

Let's review a few types of flight modes that GPS allows.

Loiter (position and altitude)

Probably the most-used GPS feature, *loiter* does exactly what the name implies: it holds the aircraft at a specific location (latitude, longitude, and altitude) at the flick of a switch. When in position hold, the aircraft will self-correct its location if it is changed by any outside forces. For example, once loiter is engaged, the aircraft would automatically fly back to its original location if the wind pushed it away. This is typically within a range of plus or minus two meters when properly configured. The same approach is taken with altitude. Altitude hold is usually within plus or minus three meters, and can use additional sensors beyond just GPS, such as the internal barometer.

Return to Home (RTH) failsafe

One of the best safety features in modern UAV—Return to Home (RTH)—is only possible when GPS is used. This mode allows the aircraft to determine a home location, usually the original take-off location, and return there under certain failsafe circumstances. Losing radio connection to your aircraft is one common trigger for RTH. This could happen if your radio batteries die while in flight or if you fly outside of the radio's broadcasting range. Some people also program a dedicated switch on their radio to engage RTH.

RTH Safety

While this is a great feature that can surely save your aircraft in a time of need, it should only be used as a failsafe, not a common feature. Never fly your aircraft so far away that you cannot bring it back on your own. In other words, do not *rely* on the RTH feature.

Waypoint navigation

A slightly more advanced use for GPS data is autonomous navigation through programmed waypoints. Ground control software uploads a list of navigation instructions to the flight controller, which carries them out, step by step, as a complete mission. This technique is especially useful in the UAV mapping industry, where specific flight patterns need to be flown time and time again. Always confirm optimal GPS performance on your aircraft before attempting to fly via waypoints. If the craft can hold position and RTH safely, it should be more than capable of waypoint flight.

Is My GPS Working Correctly?

One nice way to tell how accurate your GPS is without running the risk of complicated flight mission is to watch its location inside your ground control software. Place the aircraft in one stable position in a big open field without moving it. If the aircraft appears to be drifting in the ground control software, you may have some interference affecting your GPS. Try repositioning it and repeating the test. Another option is to raise it higher away from other electronic components.

Follow Me mode

The newest use for GPS data in the multirotor world is currently *Follow Me functionality*. This mode allows a user to send a real-time stream of the user's location data to the drone, through the use of a mobile app. The drone then uses that location data from your mobile to follow you from a predetermined height and distance. The possibilities for sports coverage alone are fantastic—imagine a camera drone always hovering 20 feet above the quarterback's head—but Follow Me should be used with caution. Because most small UAVs do not have Sense and Avoid technology, Follow Me mode could cause trouble in tight or congested environments. If you ski under that tree branch, your quad won't be able to tell it's there and could fly right into it.

Compass

While GPS allows our drone to obtain a higher level of autonomy, it wouldn't be worth very much if it couldn't keep its heading. That is where the compass comes into play. Often bundled together with the GPS into one unit (as in our demo build), the compass tells our drone not only where all the cardinal directions lie, but also which direction it is pointing to upon startup and any changes that happen during its flight.

Is It Really a Compass?

For most of us, the term compass probably conjures up images of that shiny round instrument you saw on your first hiking trip or scouting adventure. You know the one I'm talking about, with the needle that always points north. Your APM does not actually have such a device built inside of it. It uses a small electronic sensor called a magnetometer to find its compass bearings. Just like a traditional compass, a magnetometer measures the magnetic field in its immediate surroundings and uses that information to determine which direction it is facing—only it does so without the fancy needle. For the duration of this book, the terms compass and magnetometer will be interchangeable.

APM 2.5 Versus 2.6

Depending on how you sourced the parts for your build, you are likely to end up with one of two different versions of the APM flight controller. There are a few subtle differences between the two, but the main thing we are concerned with is the compass. The APM 2.5 used an internal compass while the 2.6 switched to an external compass that is bundled inside the GPS housing. This allowed the sensor to be removed from the inside of the APM housing and placed in a much more suitable environment (magnetically speaking) such as the GPS enclosure. If you have an APM 2.5, don't fret! There is a fairly simple way to disable the internal compass and take advantage of the newer external units (*http://bit.ly/ ext_compass_install*).

Battery Monitor

Another small but very important accessory in our build is the *battery monitor* (also called the *power module* in some circles). While some other flight controllers have some form of battery monitoring built in by default—such as an internal power module—APM allows you to add it as an optional sensor. Although it's optional, we highly recommend that you add one to your aircraft (see Figure 5-2). It provides some of the most important data you could ask for and generally comes with most APMs as part of the basic bundle (depending where you buy it from).

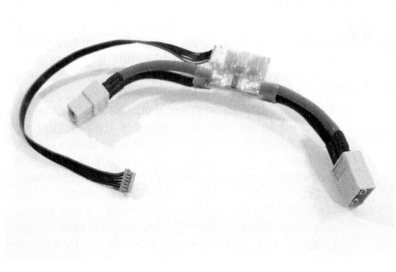

Figure 5-2. *Our inline battery monitor collects very important mission data as we fly.*

The theory behind the battery monitor is simple: place a small circuit board inline with the main battery lead and allow it to collect critical voltage and amperage information that is fed back into the APM. From there, we can direct the firmware to do a number of different things with that information, such as automatically return to home (RTH) when a certain voltage is reached as a failsafe measure.

Step-by-Step Build Instructions

The following steps will walk you through the installation of the GPS, compass, and battery monitor. While this portion of the build may seem fairly simple from an install perspective, please pay careful attention, because getting these wrong could have a huge impact on your aircraft stability and reliability:

Step 1: Mount the GPS Puck

Installing the GPS puck is probably one of the simplest steps of this entire build. It is important to know that most GPS pucks also contain a compass, which means they have a dedicated front and back. Always make sure that the forward arrow is pointing toward the front of the aircraft. If your GPS component came with a raised mounting bracket, use it (see Figure 5-3). If it didn't—and you don't have any way of making one—you can order one from 3D Robotics or the Maker Shed. It is generally considered a best practice to place the GPS puck (with internal compass) high above the rest of the electronics to avoid any magnetic interference. Follow any assembly instructions that came with your mounting bracket, then attach the bracket to the top of the drone frame as close to the center of gravity as you can. Apply a generous amount of double-sided tape to the bracket—or find a way to screw it in place—in order to hold it in place.

Figure 5-3. *3DR makes a nice mast for the uBlox GPS.*

For our build, we made our own GPS mount out of an extra 37-mm standoff and a 3D-printed base (see Figure 5-4). That is the unit you will see in most of the remaining figures in this chapter. Depending on which GPS unit you are using, several open designs (*http://gettingstartedwithdrones.com/gps/*) are available on the Internet for 3D-printable mounts.

Figure 5-4. *We like to keep things simple with this DIY GPS mast.*

Step 2: Connect GPS and Compass to the APM

If you bought the 3DR uBlox GPS for your build, it should have come with two (possibly more) small cables that allow you to connect it to the APM. Remove them from the bag and inspect them carefully. This GPS is able to work with a few different autopilot systems, so it will likely ship with a variety of cables to cover all the bases. The two that you want to find are identified as a 4-position cable and a 5-to-6-position cable. This may sound a little funny at first, but it's pretty simple. The positions indicate how many wires are able to be fitted into the end of the plug. In our case, we have one cable that has four positions on both ends with four wires that run between. This one is for the compass (magnetometer). Our second cable has a 5-position plug on one end and a 6-position plug on the other with five wires that run between them. That's right, one of your plugs will have an empty spot in the plug; this is normal, so do not be alarmed. We will get into why that is the case later, but for now, have a look at the cables we will be using in Figure 5-5. If you did not get these cables with your compass, tell your vendor immediately, especially if you bought the APM and GPS together from the same store!

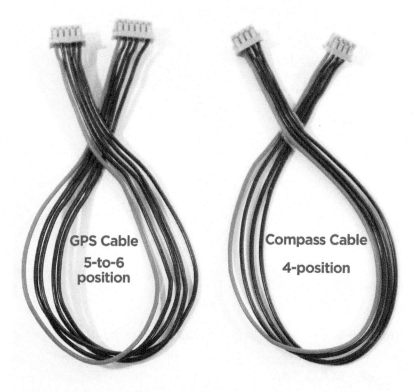

Figure 5-5. *Make sure that you have the right cables for the job.*

Now that we have identified the proper cables, let's take a look at the GPS unit itself. There is not much to it; it's a small black square with two ports on the left side labeled GPS and MAG. Upon closer examination you will notice that the GPS port has six small pins inside it and the MAG port has four. Is this starting to make sense now? The 5-to-6-position GPS cable will plug the 6-position end into the GPS port, while the compass cable will take either end of our 4-position cable (see Figure 5-6).

Figure 5-6. *The two ports on the side of our GPS puck.*

 MAG Port

In case you haven't figured it out on your own, the MAG port on the side of the GPS puck is labeled as such because it is short for magnetometer. This is the technical name of the sensor that provides our compass functionality.

Now that we have one end of our cables plugged into the GPS puck, let's plug the other ends into the APM. Find the 5-position plug that is connected to the GPS port (on our GPS puck) and plug it into the GPS port on the APM (next to the forward arrow on the top). Now, take the remaining 4-position plug and insert it into the I2C port (short for inter-integrated circuit and pronounced I-squared-C). Refer to Figure 5-7 for details on how these two cables are connected.

Figure 5-7. *Our GPS and compass (magnetometer) plugged into the APM.*

Careful with the Plugs!

It can be difficult working with these types of sensor plugs. They are very delicate and can be damaged if you unseat them once they are plugged in. If you are working on your drone and have to plug and unplug these for whatever reason, take caution! We recommend using a tiny flat-head screwdriver to gently lift them out of their seat while pulling ever so slightly on the cable itself. *Do not pull too hard!* The wire will pull itself out of its plug position if you are not careful. It happened to us several time before we got used to working with these types of cables. Over time, it will become easier for you, but you should take extra caution when you first start working with them.

Which Way Does the Plug Sit?

Here is a quick and easy way to tell how the plug should be inserted into your component. All cables being inserted into the APM unit itself will always have the single red wire on the right-hand side (as determined by the ports text label). Oddly enough, we found the exact opposite on the uBlox GPS unit; the red wire was always on the left. This becomes apparent from where the GPS and APM are wired up in Figure 5-7.

Step 3: Install the Battery Monitor

Finally, we will install our battery monitor. Remember, this is an optional accessory. If you are choosing to leave yours off— maybe you want to save a few grams of weight—then skip this step. That being said, there is a lot of value in having this important data monitored during the flight of your drone, and we encourage you to go forward with this part of the build.

The overview of the install is quite simple: first, we plug our main battery lead from the power distribution board (or power harness) into our battery monitor, then we plug the 6-position plug into the PM port on the APM. The flight battery will now plug into the battery monitor rather than the main power lead coming off the power distribution board. Figure 5-8 shows the battery monitor plugged in on the bench.

Figure 5-8. *Our battery monitor plugged in on the bench, speaking to the APM over a 6-position cable that plugs into the PM port.*

Start off by plugging the battery monitor into the main battery lead from the power distribution board (see Figure 5-9). This is the same lead that you would normally plug your battery into if you weren't using a battery monitor.

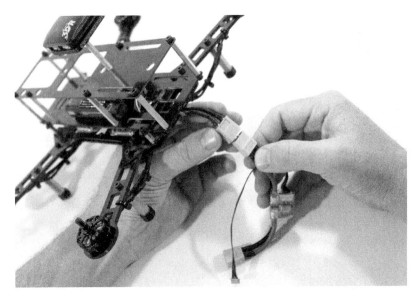

Figure 5-9. *Your battery monitor should have the same type of plugs as your main battery lead (XT-60, in our case).*

Next, we need to mount our battery monitor on the back portion of our frame (see Figure 5-10). Start by pushing the original battery lead plug into the space between the clean and dirty frames all the way in the back of the frame. It should fit in there nice and snug with the monitor lying across the bottom of the clean frame. Be sure to route the 6-position cable across the bottom of the frame in through the same holes that you routed the ESC plugs earlier in the book. This cable will be long enough to reach the PM port on the APM, its final destination.

Because the leads on our battery monitor are so short, we can attach the monitor itself to the back end of the frame with a zip tie or two leaving the new battery lead exposed and easily accessible to our flight battery (see Figure 5-11).

Figure 5-10. *Getting our battery monitor into position and ready to be fastened to the frame.*

Figure 5-11. *The new battery lead is accessible to the flight battery.*

Now there's only one thing left to do: plug the 6-position cable that comes out of your battery monitor into the PM port as shown earlier in the chapter. Make sure that the red wire within the cable is on the right side (as indicated by the port label).

Wrapping Up

That's it for the GPS, compass, and battery monitor. Later in the book, we will use the flight controller software to confirm that the GPS is picking up the signal that we are looking for. We will also take a look at the settings for the battery monitor. For now, move on to the next chapter.

6/Transmitter

What Is a Transmitter?

So far we've talked about a lot of different components of your aircraft but haven't talked about how the aircraft is controlled. That is where the *transmitter* comes in (see Figure 6-1). The transmitter is the remote control for your aircraft. You can use the transmitter to send commands to your aircraft by manipulating any of its sticks, sliders, buttons, or switches. The commands are sent from the transmitter to a receiver on the aircraft, which sends those commands to the autopilot or other components.

Figure 6-1. *The popular Taranis transmitter, made by FrSky.*

Most Common Frequency Bands

In the early days of RC aircraft, modelers operated in the lower range of the AM frequency band. This works great for long distances but has one major drawback: only one person could be tuned to that frequency at a time. If your friend had the same channel set up on his transmitter, he could interfere with your aircraft just by turning his on.

Modern-day transmitters have overcome this problem by moving to the higher 2.4 GHz frequency band and employing frequency hopping to ensure that no one is ever on the same frequency at the exact same time. Even more recently, some manufacturers have begun to move their RC control channels up to the 5.8 GHz frequency band. This is done so that they can reserve the 2.4 GHz frequency band for video transmission (which we will discuss in greater detail in Chapter 8). It is important to know exactly which frequency bands you are operating on, because flying in groups where conflicting frequencies are duking it out for airspace can cause trouble.

 A Good Rule of Thumb
Try not to use 5.8 GHz control transmitters with 5.8 GHz analog video transmitters. By the same token, you should keep 2.4 GHz control transmitters away from 2.4 GHz analog video transmitters. These components can cause interference if they are on the same frequency. Instead, use 5.8GHz controllers with 2.4GHz video transmitters, or vice versa.

Different Modes Around the World

One important thing to know when buying an RC transmitter is what mode it operates on. Mode 2 is used in the United States. This places the throttle and rudder on the left stick, with the pitch and roll on the right stick. Mode 1 is used throughout Europe and Japan and reverses the two sticks so that the throttle is on the right.

PWM Versus PPM

If you have shopped around for RC transmitters before, you have likely seen the terms PWM and PPM associated with them. They stand for *pulse-width modulation (PWM)* and *pulse-position modulation (PPM)*, respectively. These labels identify different communication protocols between your receiver (see Figure 6-2) and the components on your aircraft. The main technical difference between the two is that PPM is digital while PWM is analog. As a builder, you will notice that the biggest install difference between these two is that PPM communicates up to nine channels of control across one wire (digital serial communication), while PWM requires a single wire for each individual channel (analog communication). It is important to consult the documentation for both your receiver and autopilot to know which protocol you should use. If PPM is available, we highly suggest that you use it, because it will clean up your install dramatically with fewer wires. However, not all hardware supports the PPM protocol, so consult your owner's manual for further reference. At the time of writing this book, APM does not support digital communications protocols on its main firmware fork. As far as we can tell, some developers have been able to create custom firmware that supports it, but we do not advise you to install untested firmware from indie developers as a newcomer to this hobby. Stick with the official releases for now, and you will be much happier in the short term!

✎ Digital by Any Other Name

Many manufacturers have developed their own proprietary versions of digital serial communications and give them their own names. For example, Futaba calls its version SBus.

Figure 6-2. *Six-channel receiver by Spektrum.*

Step-by-Step Build Instructions

For this portion of the build, you will need:

- One inch of foam double sided tape
- Scissors
- Male-to-male servo leads

Step 1: Identify What You Need

The high-level idea behind installing the RC receiver is easy. You physically mount the receiver on the frame, then plug a series of cables in between the receiver and the autopilot. Because of the differences between PWM and PPM, as well as different manufacturer standards, there are dozens of different connection scenarios between your receiver and autopilot; however, we will only be covering one here. That example will be a PWM connection to a Spektrum style receiver (see Figure 6-3). If you are building something different, consult the owner's manual of both your autopilot and transmitter for the correct connection sequence.

Figure 6-3. *Our Spektrum receiver is plugged into the APM input on the bench.*

 Which Way Is Front?

Be sure that you are mounting these components in the correct spot. You can easily identify the front of the frame, because it has a large cut out for the vibration isolation mount that the camera will sit on. Another good reminder is that the booms fold back on the frame, not forward.

Step 2: Mount the Receiver

Use a piece of the thick foam double-sided tape to mount the receiver somewhere on your airframe (see Figure 6-4). We recommend placing it inside the front of the frame, as this will give you easier access to the input ports of the APM later in the build.

Keep a Clear Line of Communication

If you are building a kit other than the Little Dipper, try to find a location for your receiver that is close to your autopilot and makes it easy to maintain a clear line of sight with your controller on the ground. Remember that some materials such as carbon fiber can block radio transmissions. It's a good idea to keep as little airframe as possible between the receiver and the controller on the ground. If your receiver has long antenna wires, run them out of the frame and as low as possible to provide the best possible reception.

Figure 6-4. *Spektrum Receiver installed in the nose of the airframe.*

Step 3: Plug in the Receiver

Take five short male-to-male servo wires and plug them in according to Figure 6-5. Connect input 1 to your receiver's roll channel, input 2 to the pitch channel, input 3 to the throttle channel, input 4 to the yaw channel, and finally input 5 to Aux 1. We will use this channel to switch among our different flight modes.

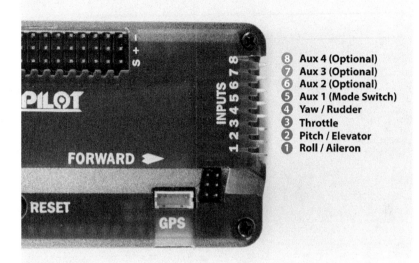

Figure 6-5. *PWM wiring schematic for APM.*

7/Telemetry Radios

Software Monitoring and Control

By now, you may have noticed one very important thing about RC transmitters: they only send communications in one direction, from the transmitter to the aircraft. In order to gain access to any of the onboard data from your aircraft, we have to use a special two-way radio that is known as a *telemetry link*.

Pick the Right Frequency

Telemetry radios operate on two standard frequencies. In the United States, we typically use the 915 MHz band, while the 433 MHz band is favored in Europe. These standards are established to help avoid interference from other electronics on the open market. Operating on the wrong frequency band can be illegal in certain parts of the world. Be sure to check the laws in your country to see what standard you should use.

Just as its name implies, the telemetry link takes a series of measurements from your aircraft and sends them back down to the ground where they can be displayed in ground control software (Figure 7-1). In addition, data—such as waypoints, autonomous flight missions, and aircraft configurations—can be communicated up to the aircraft from the same ground control software.

Having access to this type of information while you are still safely on the ground can be a huge advantage. It can help you identify problems before they get out of hand. Some of the data points your ground control software needs are latitude, longitude, altitude, aircraft attitude, current battery voltage, aircraft heading, speed and duration of flight, plus many more. You can find a complete list of all data communicated in your ground control software manual.

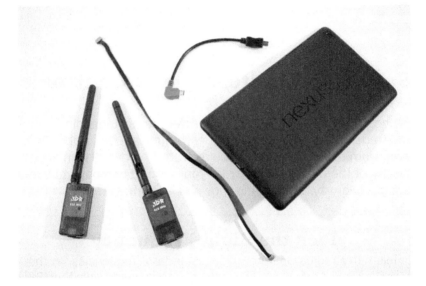

Figure 7-1. *This Android tablet can be used a ground station with the addition of telemetry radios and a free app.*

Step-by-Step Build Instructions

For this portion of the build, you will need:

- One inch of foam double sided tape
- One inch of Velcro
- Zip ties

Step 1: Mount the Ground Station Radio

Your telemetry radio set will actually come with two different radios. One goes on the aircraft while the other connects to your ground control software (Figure 7-2). The ground radio can be mounted in any number of ways; however, we ordinarily use Velcro on the back of the tablet or laptop that runs the software. This allows us to remove the radio when we're not using the ground control software.

Figure 7-2. *Velcro holds the telemetry radio on the back of our tablet while in use and allows for easy removal when we are not flying.*

Ground Station Software

Because APM is an open source platform, a number of ground station software bundles (*http://bit.ly/ground_stations*) are available for it. Our personal favorites are Tower for the Android tablets and Mission Planner for Windows-based PCs. If you are using a Mac laptop, check out APM Planner. It is very similar to Mission Planner, but it's cross-platform with builds available for Windows, OS X, and Linux.

Step 2: Prep the Aircraft Radio for Mounting

For the aircraft radio, we suggest using double-sided foam tape to attach the radio to the frame (Figure 7-3). Cut two small strips about 1/2 inch in width and place them on opposite ends of the radio. As for location, there is some flexibility here, but we recommend you place the radio on the back half of the airframe.

This will move the telemetry radio as far away from the RC receiver as possible.

Don't attach the radio just yet. We want to put the tape in place and have it ready, but we still need to plug the radio in, and this will be much easier to do if it's not mounted yet.

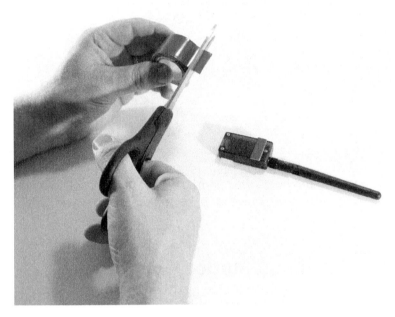

Figure 7-3. *Double-sided foam tape can be used to fasten the radio to the air frame.*

 ## Which Way Is Up?

Be sure that you are mounting these components in the correct spot. You can easily identify the front of the frame because it has a large cut out for the vibration isolation mount that the camera will sit on. Another good reminder is that the booms fold back on the frame, not forward.

Step 3: Plug in the Radio

Wiring up your new telemetry radio couldn't be simpler (see Figure 7-4). Find the correct cable that came with the radio set and plug it into the back half of the aircraft radio. This cable will have five individual wires (four black and one red), but the plug on either end will have slots for six wires, leaving one of them empty. This is normal. The other end of that cable will plug into the telemetry port on your autopilot.

Figure 7-4. *Plugging in the telemetry radio will be much easier if done before it's mounted to the frame.*

Step 4: Mount the Aircraft Radio

Now that you have plugged in the radio, go ahead and mount it on the back of the airframe (see Figure 7-5). The double-sided tape should hold it securely in place. Make sure that it sticks out from the frame far enough to allow the antenna to swivel freely as needed, but not so far that it will be in the way as you handle it.

Figure 7-5. *The end product with the radio mounted in place.*

You will notice small tabs on the sides of the cable plugs that allow the cable to plug in in only one direction. Find those tabs and identify the direction before applying pressure (see Figure 7-6).

Figure 7-6. *Guide tabs on the side of each plug allow them to only be plugged in correctly.*

8/Camera and FPV Equipment

Aerial photography is currently drones' number one use, so the topic of cameras on drones could be a book all its own. That book could cover a wide spectrum of setups, from large $50,000 Hollywood cinema rigs, all the way down to small action sports cameras, such as those that we will be using on our demo build. This chapter touches on the basics of having a fixed-mount camera with the ability to broadcast live, usually to a small monitor or video goggles.

Popular Drone Cameras

Selecting the right camera for your drone can feel like a difficult task if you're new to the topic. The top five things that we always look for when selecting a new camera for any of our drones are:

Weight and size
 As with every other component in the drone world, weight is always one of the first qualities to consider. It is always more desirable to have a lighter component.

Remote trigger
 When shooting video, you can simply start the video recording while on the ground and stop it when you land. However, it is a little more complicated for still photos. Still photos need to be triggered continuously throughout the flight. One popular solution is to integrate an *intervalometer* into the camera. This is basically a timer that triggers the camera shutter every *X* number of seconds. Other popular solutions include a mechanical servo arm that can physically press the shutter button on the camera, or even an infrared remote that can be controlled by a secondary RC transmitter.

Live stream

Another mission-critical feature for any drone camera is the ability to broadcast a live stream through some type of output port located on the camera. This port could be either a digital High-Definition Multimedia Interface (HDMI) output, or an analog signal through any number of different types of plugs. Believe it or not, an analog signal is the preferred method most of the time. The reason for this is that our video transmitter will only accept analog signals.

Camera resolution

Once you find a few cameras that meet the requirements just listed, you'll need to consider overall resolution. As far as video is concerned, resolution is a pretty easy nut to crack. Even small action sports cameras come in resolutions as high as 4,000 these days. Still photography, however, is another story. Always check the megapixel rating to be sure it meets your needs.

Cost

After you've whittled down the list of potential candidates using the first four qualities, it's time to look at price. Prices can range within each category, so do some research to make sure you're getting the right deal. Expect to pay up to $100 for a small keychain camera, between $100 and $500 for a decent action sports camera, somewhere between $300 and $1,000 for a small, micro four-thirds camera, and anywhere from $500 up for a digital single-lens reflex (DSLR) camera.

Micro Four Thirds Cameras

In 2002, the four thirds (4:3) completely digital, open standard system was introduced, which allows for interchangeable camera bodies and lenses from different manufacturers (see Figure 8-1). The name comes from the image sensor aspect ratio, which is 30%–40% smaller than full-size DSLRs. Yet, this sensor is still about nine times larger than a compact "point and shoot" camera.

The most obvious advantage that the four thirds systems brought to the UAV world is that a smaller sensor made it possible for smaller, more lightweight cameras to be developed. In turn, smaller, more shallow lenses made it possible for light to hit the camera's sensor in a more perpendicular direction. A direct light path increases the depth of field, which reduces the risk of out-of-focus photos. This also produces images with brighter corners and better overall off-center resolution. Basically, with a four thirds camera, you can achieve high-quality photos with a relatively small camera payload.

For our demo build, we suggest using either the Mobius action sports camera, or the 808 keychain camera.

Figure 8-1. *A very popular micro four thirds camera, the Lumix GH4 camera body with an Olympus 17-mm wide angle lens.*

The Mobius ActionCam

For our little quadcopter, we recommend using the Mobius ActionCam or something in the same weight class (see Figure 8-2). This particular model is well suited for an aircraft the size of the Little Dipper due to its light weight and ease of use. A few of its key features are:

- Small footprint: 61 mm (L) x 35 mm (W) x 18 mm (H).
- Lightweight: approximately 38 grams (depending on model purchased).
- Internal rechargeable Li-Po battery: no need to connect it to the aircraft's power system.
- Full HD video at 1080p/30FPS; 720p/60FPS for slow-motion HD video.
- Still photo intervalometer (time lapse mode) capture up to 2304 × 1536. No need for external camera trigger; it triggers itself every *x* seconds.
- Live video feed available while shooting video, excellent for hooking up to a video transmitter.

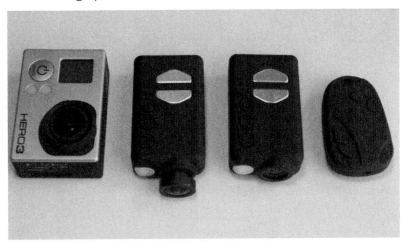

Figure 8-2. *ActionCams side by side: GoPro Hero 3, Mobius wide angle, Modius standard lens, and 808 keychain cam.*

FPV for a Live Stream

One fantastic feature to have on your drone is the ability to see what it sees in real time. This is made possible by adding a video transmitter. The transmitter, or vTX (see Figure 8-3), takes the live stream from your camera and broadcasts it over a specific frequency range to a receiver, or rTX, on the ground. The receiver then connects to a monitor or video goggles to display the live feed.

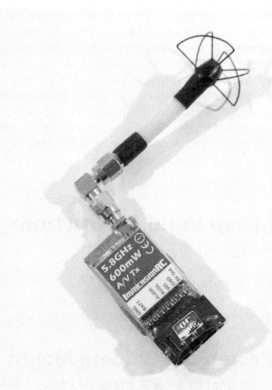

Figure 8-3. *5.8 GHz video transmitter with optional cloverleaf antenna and right angle.*

Traditionally, this communication has taken place over a 5.8 GHz analog video transmitter/receiver pair. More recently, though, some manufacturers have started to use 5.8 or 2.4 GHz WiFi networks with mobile devices to display this video stream. The overall architecture is the same for both designs: you still have a transmitter that sends the signal down to the ground where it is picked up by receiver and then displayed on some type of monitor.

What Can We Fit on Our Kit?

Because we are building the autonomous Little Dipper kit, our threshold for additional gear is very low. If you have the APM, GPS, and camera onboard, you will really only have the ability to add one more thing. In short, you have a choice between adding an FPV transmitter or telemetry radio, but we do not recommend adding both. We have already walked through the process of adding the telemetry radio earlier in the book, so we are going to continue down that build path and will not be reviewing further info on FPV. However, if you are interested in learning more about how to install a video transmitter or other FPV gear, feel free to check out our web extra on the topic (*http://bit.ly/fpv_install*).

Step-by-Step Build Instructions

For this portion of the build, you will need:

- One inch of foam double-sided tape
- Scissors
- 12 inches of thread or unwaxed dental floss

Step 1: Fasten the Camera Mount

As you remember from Chapter 2, the Little Dipper frame has a small vibration isolation plate designed to hold a small Action-Cam. If you are installing a Mobius camera on your quad, it should have come with a quick-release mount (see Figure 8-4). Our first step will be to fasten the quick-release camera mount to the camera isolation plate. To do so, simply cut two small pieces of double-sided foam tape approximately 1/2-inch long and apply it to the bottom of the quick-release plate. Now peel the protective coating from the tape and attach the quick-release mount to the vibration isolation plate (see Figure 8-5).

 ## Make Sure It's Straight!

Double-sided tape can be difficult to pull off of smaller plastic parts, such as our quick-release camera mount. Take your time and make sure that you line the camera mount up nice and straight on the vibration isolation plate. If you don't, it can be a pain to pull it apart and reapply, and no one wants a camera that looks crooked down the frame!

Figure 8-4. *Quick release mount for the Mobius camera.*

Figure 8-5. *Quick-release mount after being attached to the Vibration Isolation Plate.*

Step 2: Attach the Vibration Isolation Plate

If you haven't done so yet, now is a good time to attach the vibration isolation plate to the rubber ball mounts on the frame. A quick and easy way to do this is by using a piece of thread or unwaxed dental floss to pull the ball mounts through the mounting holes, as shown in Figure 8-6. Repeat this step for all four rubber balls.

 ### Are You Facing the Right Direction?

Make sure you have the camera mount facing in the correct direction. The mount basically has three sides with one end being open for the camera to slide into. Make sure that the open end points toward the front of the aircraft.

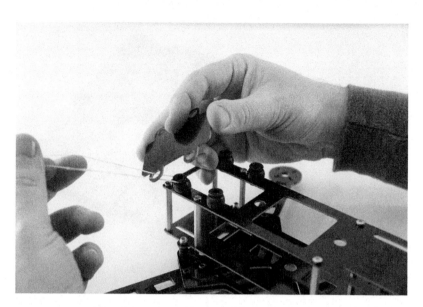

Figure 8-6. *Attaching the vibration isolation plate to the rubber ball mounts.*

Step 3: Place the Camera in the Quick-Release Mount

Now that we have our quick-release camera mount attached to the vibration isolation plate—and that plate is installed on the rubber ball mounts—all we have left to do is slide the camera into the camera mount (see Figure 8-7). Make sure that the lens of the camera is pointing away from the camera mount, then line up the small groove that runs around the middle of the camera with the five tabs that stick out from the sidewalls of the camera mount. These tabs are what hold the camera in place when it's positioned inside the mount. Now, gently push the camera back into the mount while holding the back of the mount for extra support.

Figure 8-7. *Mobius camera next to the quick-release camera mount.*

9/ArduPilot Mega (APM) Setup

At this point, the vast majority of your physical build is complete. Now it's time to jump into Mission Planner and start setting up your autopilot. Once we complete this process, we will have just a few minor things to button up on the build and then our aircraft will be ready to fly!

This chapter will be structured a little different than the rest, as we will go straight into the step-by-step instructions before looking at the rest of the Mission Planner application.

Step-by-Step Build Instructions

For this portion of the build, you will need:

- A laptop PC running Windows 7 or newer (if you only have a Mac, try setting up a Boot Camp partition to install Windows)
- The most recent version of the Mission Planner application
- A Micro "B" USB 1.1/2.0 cable

Step 1: Update Firmware

The very first thing we should do is make sure we are running the most current version of the firmware. Depending on when your APM was made, it could have a number of different versions on it. Don't take a chance using what it shipped with; go through the process of updating, and then you can be sure you are on the right track:

a. Power up your laptop and open Mission Planner.
b. Plug the USB cable into the APM with the other end going into your laptop.

c. Go to the Initial Setup tab in Mission Planner and click Install Firmware in the left-hand menu.

d. From the graphic buttons in the middle of the screen, select ArduCopter vX.X Quad (where vX.X is your version number) as the firmware type; see Figure 9-1.

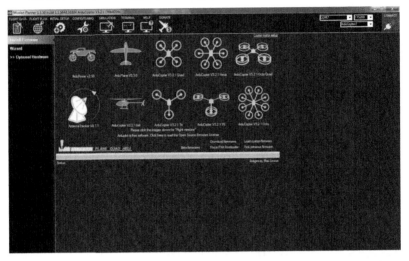

Figure 9-1. *Select the type of firmware you need to update (ArduCopter Quad, in our case).*

e. You will be asked to confirm that you do want to update the firmware. Click Yes to continue (see Figure 9-2).

f. Mission Planner will locate the most recent version of the APM firmware that is available for your hardware and start the update process. Click OK to proceed (see Figure 9-3).

g. Firmware uploading begins (see Figure 9-4).

h. After the firmware is uploaded, it needs to be verified (see Figure 9-5).

i. After the update is complete, Mission Planner alerts you to any additional steps that may need to take place as a result of the update (see Figure 9-6).

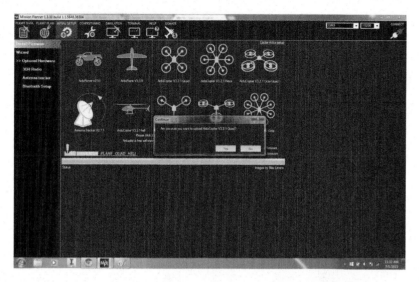

Figure 9-2. *Confirm that you want to upgrade by clicking YES.*

Figure 9-3. *Mission Planner automatically finds the most recent firmware version supported by your APM.*

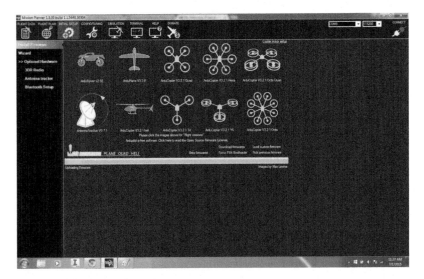

Figure 9-4. *Firmware upload has begun.*

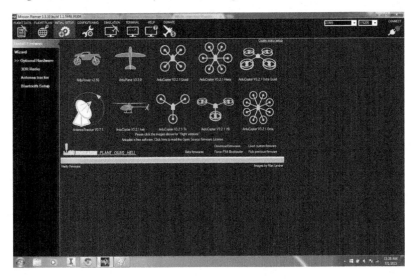

Figure 9-5. *Firmware verification in progress.*

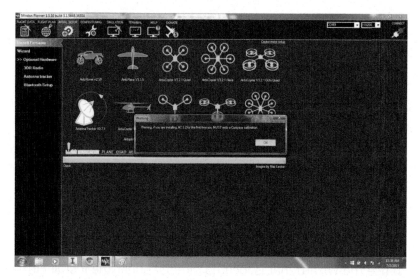

Figure 9-6. *We are all done! Click OK to proceed.*

Stay Connected

Do not disconnect the USB cable from your APM until the update is complete. It's very important that you keep an eye on the progress bar to see when the upgrade is complete and never interrupt the process.

Baud Rates and COM Ports

Any time you are connecting to your APM with Mission Planner, you must indicate the proper baud rate at which your communications should travel. This defines the bits per second for data communication. The general rule of thumb is to use 115200 for communications over a USB cable and 57600 for communications over the telemetry radio. Other applications may detect the communication type and adjust this for you automatically, but it's still a good thing to know, because Mission Planner asks for this input.

In addition, you will need to know on which COM port your APM is communicating with your laptop. Mission Planner is pretty good about finding your APM and selecting the correct COM port by default, but there is a trick to that. You need to plug the APM in before you open Mission Planner. If this still doesn't work, look into your PC settings and find all open COM ports. You should see your APM listed there as an Arduino Mega. Take note of that COM port and remember it moving forward.

Step 2: Connect and Complete Mandatory Setup

Now that we have updated our firmware, we have a nice clean slate to start from. Let's walk through our mandatory setup now and get this bird ready to fly!

There are actually two ways you can complete this step. One is to use the Wizard that is located directly under the Install Firmware button you used in the last step. The second way is to do it manually. We are going to walk through the manual process in this example, but both ways are essentially the same thing; the wizard just forces you to do all the steps together, while manual entry allows you to do only the steps you want. Any time we set up a new copter, we want to complete all of these steps, but there may be times in the future where you also want to revisit just one or two of these steps—such as GPS or IMU calibration —and we want to you show you that from the beginning.

Before we can set anything up, we need to connect our APM to Mission Planner. This requires you to select the COM port and Baud Rate before clicking the Connect button in the upper-right corner of the application. Once you connect, you will notice a new set of menu options appear in the left-hand column, including Mandatory Hardware and Optional Hardware.

Step 2.1: Set frame type

Click the Frame type button from the left-hand column under the Mandatory Hardware heading and select the X copter category (see Figure 9-7).

💣 Ignore the Default Settings

You will notice that next to your frame selections is a drop-down menu titled Default Settings. You can ignore that during this build. It is used as a shortcut for certain off-the-shelf drones. For example, if you owned a 3D Robotics Iris+ quad, you would find settings in this menu geared specifically toward that model. Unfortunately, Mission Planner doesn't come bundled with default settings for our model, so we have to set them up manually. That's OK; you learn more doing it that way!

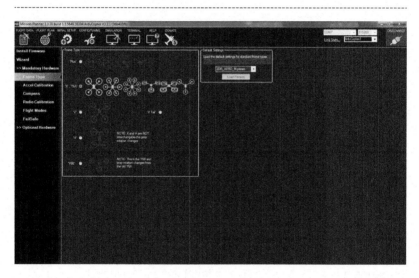

Figure 9-7. *Select frame type.*

X Versus +

When we selected the X frame type, you probably noticed that a number of options are available. The most common are X and + (sometimes called I by other software). This refers to the configuration of the frame and how it relates to the positioning of the APM. If you can imagine a straight line that runs through the length of your APM and extends out the front, the X configuration would point that line between two motors, whereas the +

configuration would point directly toward one. Figure 9-8 illustrates this.

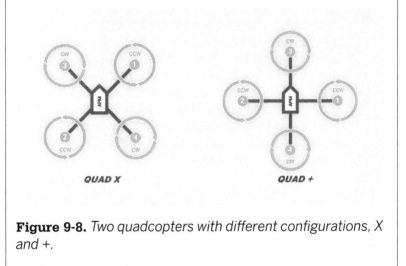

Figure 9-8. *Two quadcopters with different configurations, X and +.*

Step 2.2: Calibrate accelerometer

The next substep is to calibrate our accelerometer. This sets a baseline for the APM to understand what direction is up, down, and so forth. Mission Planner will ask you to move the drone into different positions, so be sure you have a nice, clear area on a level work surface to follow the directions as described. It is important that you do this on an area that is as level as possible. If you have one, use a bubble level to confirm that the area you have selected is not out of level. We often use a small TV table because it is easy to shim up to level if it's off a little bit.

Click the Accel Calibration button from the left-hand menu (see Figure 9-9). This will place a single button labeled Calibrate Accel in the main application window. Click that button and follow the instructions that Mission Planner sets before you. It will ask you to position your drone in a number of different positions before pressing any key to continue. Once the process is done, you will see a notification that the calibration either passed or failed. The positions it will ask you to use are:

Level

Normal position the aircraft would maintain when sitting right side up.

Upside down

Exactly what it sounds like: flip the aircraft upside down from how it would normally sit.

Nose down

The front of the drone pointed straight down toward the ground with the tail sticking straight up.

Nose up

The tail of the aircraft pointed straight down with the nose pointed straight up.

Left

The left side of the aircraft pointing straight down with the right side pointed straight up.

Right

The right side of the aircraft pointed straight down with the left side point straight up.

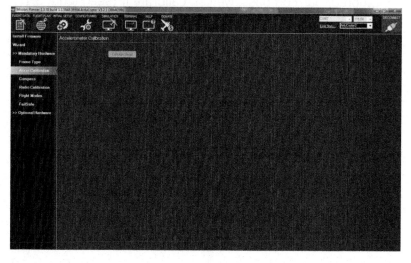

Figure 9-9. *Calibrate accelerometer.*

Step 2.3: Set up the compass

Click the Compass button in the left-hand menu. This will present the compass setup screen in the main application window (see Figure 9-10). Setting up the compass is a very important process that involves three basic steps:

a. *Enable auto declination:* Find and check the Enable and Auto Dec Declination checkboxes at the top of the main application window. This allows the APM to use its GPS to find the part of the world you are located in and then fine-tune your compass for that location. This allows for offsets between true north and magnetic north.

b. *Select type of compass and orientation:* Under the Orientation field in the main application window, select the APM with External Compass option. This should change the text at the bottom of that field to say ROTATION_ROLL_180. This information tells the APM what direction the compass is pointing in, so it has a baseline to start its measurements from.

c. *Perform calibration:* Now that we have established the orientation of our compass, we can calibrate it (see Figure 9-11). Click the Live Calibration button at the bottom of the main application window and follow the instructions Mission Planner issues. This step may take a few minutes as the application asks you to rotate the aircraft around all possible axes so that it can take sample points in a virtual sphere. Just take your time and follow the directions until Mission Planner tells you that the calibration has passed. If, for some odd reason, it fails, don't worry about it. Just run the calibration again and be sure to move around all axes of the compass.

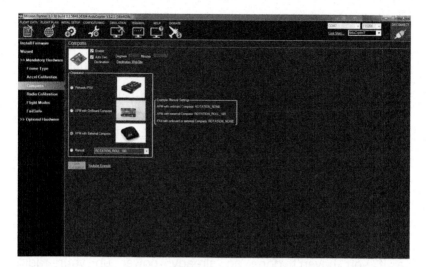

Figure 9-10. *Compass setup.*

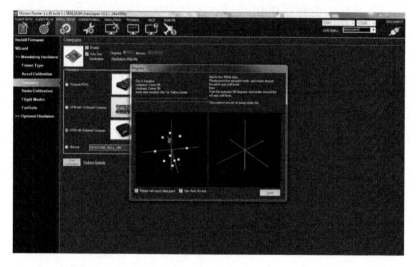

Figure 9-11. *Compass calibration.*

Step 2.4: Radio calibration

Next up is the radio calibration. APM refers to the transmitter as a radio, so from here on we use the two terms interchangeably. Click the Radio Calibration button in the left-hand menu (see Figure 9-12). This will display the calibration page in the main application window. The general idea here is to allow Mission Planner to talk with your radio and figure out the limits of each channel. During calibration you will be asked to push the sticks, switches, and knobs to their minimum and maximum settings so that the application can learn the range of each channel. A green bar graph will represent each channel of the radio, and they will move with the controls of the radio (see Figure 9-13):

a. If you haven't done so already, power up your radio transmitter.

b. Click the green Calibrate Radio button under the Radio 8 bar graph in the main application window.

c. When directed, move all radio controls to their maximum and minimum settings. This includes any and all controls that you will use such as sticks, switches, knobs, and sliders (depending on what radio you own) that are on the transmitter. You will notice that fine red lines will appear on the bar graph for each channel showing its min-max setting.

d. Once this is complete, click the Click when Done button (same as the calibrate button, the label just changed).

After you have completed the calibration, a small pop-up will appear that gives you the min-max values for each channel of the radio (see Figure 9-14). Click OK to get rid of this screen and move on to the next step.

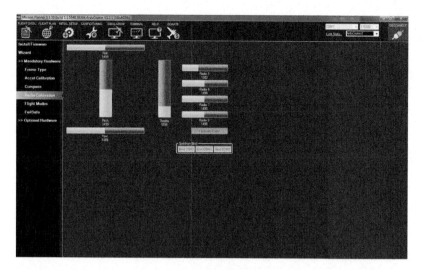

Figure 9-12. *Prior to starting the Radio Calibration process.*

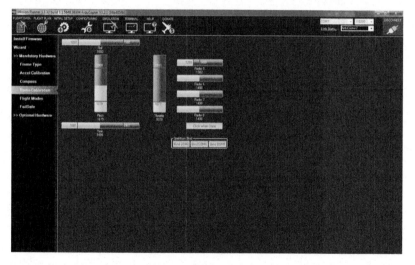

Figure 9-13. *Notice the red lines that indicate the min-max value for each channel.*

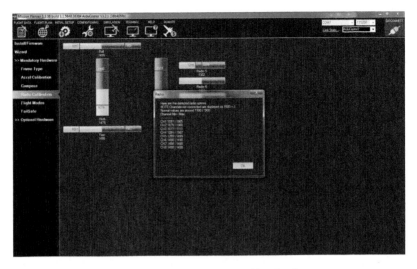

Figure 9-14. *Congrats! Your radio is calibrated.*

Step 2.5: Set flight modes

Now that we have our radio calibrated, let's set the appropriate flight modes on our flight mode switch. To start this process, click the Flight Modes button in the left-hand menu. This will display the flight mode assignment view in the main application window. We will use this to set our Loiter and Stabilize flight modes. If you have a radio that has a three-position switch (some only have two positions), then you can select one additional flight mode such as Auto (waypoint navigation) or RTH (return to home):

a. With the radio powered on, toggle through the different positions on your flight mode switch while watching Mission Planner. You will notice that each position on your switch will light up a different flight mode menu green in the application. Make note of which switch position lights up which mode menu.

b. Use the drop down menu for each flight mode to assign the desired mode to that switch position. For example, if you have position 1 on your switch lighting up mode menu 2 in the application, you can assign the Loiter flight mode to that switch position by simply changing it in the drop-down menu (see Figure 9-15). Now change to switch position 2 and assign Stabilize with that mode menu (see Figures 9-16 and 9-17).

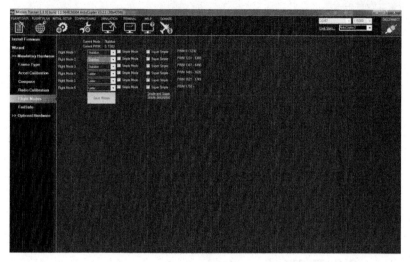

Figure 9-15. *Switch position 1 lighting up flight mode menu 2.*

Figure 9-16. *Switch position 2 lighting up flight mode menu 4.*

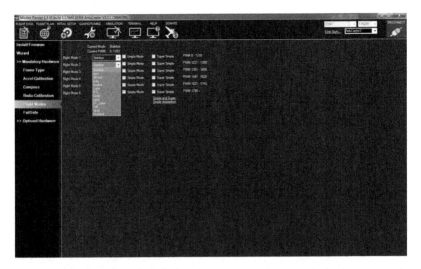

Figure 9-17. *Assigning the Stabilize flight mode to mode slot 2 with the drop-down menu.*

Step 2.6: Failsafe setup

We are almost done with our mandatory setup. The last thing on our list is the failsafe setup. This allows us to define what actions are taken when certain portions of our aircraft fail, such as when we lose radio connection or if the battery drops below a certain level while we are still flying. To get started on this step, click on the FailSafe button in the left-hand menu:

a. Find the Battery field on the right side of the main application window. Set the Low Battery option to 10.4 and the Reserved MAH to 0. This allows us to monitor only the voltage and not the capacity of the battery. You should only set the MAH setting if you use the same size battery all the time and know exactly how much you want to "leave in the tank" before you trigger a failsafe. Setting this while using more than one battery size can cause premature failsafe triggers.

b. Now we need to set the action to take when our Battery Fail-Safe event takes place. We do this by selecting one from the drop-down menu in the same Battery field (see Figure 9-18). The options are:

- Disabled
- Land
- RTH (return to home)

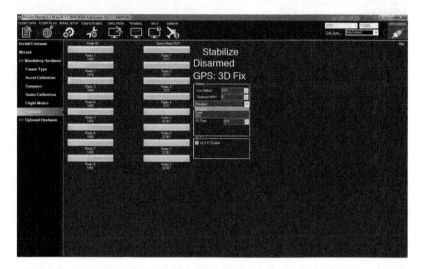

Figure 9-18. *Battery FailSafe options.*

c. In the Radio field just below, select your desired action from
the drop-down menu (see Figure 9-19). This will tell the APM
what to do any time the radio connection is dropped for a
certain amount of time (as decided by APM). Now set the FS
PWM text field to a value of 975. This tells APM what the
PWM value should be during the RTH event. The options for
the Radio FailSafe actions are:

- Disabled
- Enabled always RTH
- Enabled Continue with Mission in Auto Mode
- Enabled always Land

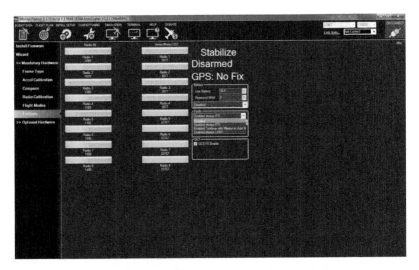

Figure 9-19. *Radio FailSafe options.*

Step 3: Optional Hardware

You have made it through the mandatory hardware setup! Now it's time to move on to the optional hardware. APM allows for a lot of optional hardware to be added on to the autopilot. This is one of the great things about the system and what attracts so many DIYers and tinkerers. The most common add-ons are telemetry radios and battery monitors, but there is a long list of additional options including sonar sensors, airspeed monitors, optical flow sensors, on screen display (OSD), camera gimbals (servo based), antenna trackers, and Bluetooth modules. Most of these add-ons fall outside of the scope of this book. In reality, you could just about write an entire book on these options alone! In this particular step, we will cover the two most popular options: telemetry radios and battery monitors. If you have elected not to install these components, feel free to skip over all of Step 3.

Step 3.1: 3DR radio (telemetry)

In Chapter 7, we installed our telemetry radios for APM. These radios come from the factory paired up in matching sets that are both tuned to the same channel, or Net ID. In theory, you should not have to do anything beyond plug these radios in and hit the connect button. However, it is good to know where the settings for the radio are located and how to change them if need be (see Figure 9-20). It is also a good idea to tune in your own Net ID if you fly with friends so that you aren't interfering with one another.

Follow these steps to check that your radios are communicating with each other as expected:

a. Make sure that you are not already connected to Mission Planner over USB. If so, disconnect and unplug the cable.
b. Power up your radio transmitter and drone with telemetry radio installed.
c. Connect the ground telemetry radio to your laptop and open Mission Planner.
d. Set your COM port as needed and your Baud Rate to 57600, then click connect.

If you want to change your Net ID setting so that you can fly with others and not cause interference, simply use the drop-down menu to find a new channel. Be sure to set both the Local (ground) and Remote (air) radios to the same ID number, then click the green Save Settings button in the top of the main application window.

Figure 9-20. *Telemetry radio setting loaded.*

Step 3.2: Battery monitor

Another optional piece of hardware that many people like to use is a battery monitor. This is a simple device but one that can provide very valuable information and potentially save your aircraft in the event of a battery failure. As we covered in Chapter 5, the battery monitor installs inline between your main power distribution plug and the battery itself. It uses a small chip that communicates with the APM to analyze the current draw and capacity remaining in the battery. This page in Mission Planner allows you to adjust the few settings for the battery monitor. To get to this screen, select Battery Monitor in the left-hand menu. There are three settings that you have control over:

Monitor

This dictates what the battery monitor will keep an eye on. Your options are Disabled, Battery Volts, and Voltage and Current. We recommend using Voltage and Current if you are using the battery monitor (see Figure 9-21). It provides more information, and that is always a good thing.

Sensor

This menu allows you to select the type of battery monitor sensor you are using. If you purchased your kit from the Maker Shed, you want to use the 3DR Power Module option. If you bought your monitor elsewhere, consult the store you purchased it from and find out exactly what type of monitor they sold you, then select that from the list. If you don't know and have no way of finding out, the Other option will usually work.

APM Version

This is pretty self-explanatory. For this build, select the APM 2.5+ 3DR Power Module option. If you purchased a different flavor of APM (such as the Pixhawk or PX4), then find your model in the options and select it.

Figure 9-21. *Battery monitor settings page.*

Let's Explore the Rest of Mission Planner

Good job—you have finished setting up your new APM equipped quadcopter! Let's take a few minutes now to look over the rest of the Mission Planner application. So far, everything that we have used was in the Initial Setup tab. We will now review some of the features in a few of the other tabs. Some will mirror features we have already discussed, but most of it will be new. Click the Config/Tuning button at the top and let's explore some of the options that appear in the left-hand menu.

Flight Modes

This screen (see Figure 9-22) is actually identical to the Flight Modes screen we discussed earlier in this chapter. You can use this page to assign the different flight modes to the various switch positions on your radio transmitter.

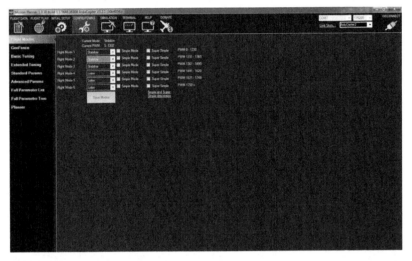

Figure 9-22. *Flight Modes screen.*

Geo Fence

Geo Fence can be a pretty cool safety feature for new pilots. It does exactly what its name implies: it sets up a virtual fence around your takeoff point that you are not allowed to fly beyond. Think of it as a way to keep your quad on a virtual leash!

APM will use your initial takeoff location to assign a home point, and from there the options you select on this screen will dictate how far away you can fly and what happens if that fence is breached. If you want to use this feature, simply check the Enable checkbox at the top of the main application window, then set your parameters below as you see fit (see Figure 9-23). Keep in mind that the parameters will be in the metric units, so the numbers indicate meters, not feet.

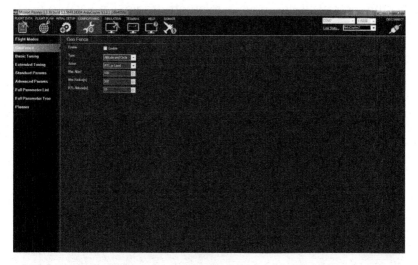

Figure 9-23. *Geo Fence screen.*

Basic Tuning

If you spend any time online reading about drones, you are likely to have heard about people tuning their copter to perform one way or another. The idea here is that any number of different drones have different flight characteristics and the autopilot— APM, in our case—needs to be tuned to match. Think of our Little Dipper; this is obviously going to have a different set of characteristics than a 35 lb Octocopter with a DSLR under it. Tuning allows you to use the same autopilot in both aircraft but alter it to work in either environment.

Basic Tuning is the entry point to tuning your APM (see Figure 9-24). One thing that attracts people to using the APM platform is that it has an almost endless set of parameters that you can use to fine-tune it for advanced functionality (most of which will fall outside of the scope of this book). This page allows you to set just a few that alter the "feel" of your aircraft. You can adjust the drone response of your input from the transmitter, how sensitive the drone is to autoleveling, where the aircraft should hover in relation to the throttle input, as well as how aggressively it will climb when given input from the throttle. Each input provides a quick description of what the slider does to the aircraft. If you have a telemetry radio, it's easy to open Mission Planner while the aircraft is in flight and slightly adjust these to see their effect in real time.

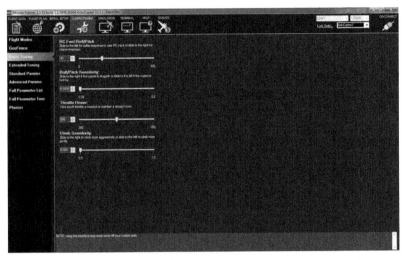

Figure 9-24. *Basic Tuning parameters.*

Extended Tuning

This Extended Tuning screen allows you to do a much higher level of flight characteristic tuning (see Figure 9-25). Sometimes called PID (Proportional, Integral, Derivative) tuning, the Extended Tuning screen lets you adjust each of the three aircraft movements (roll, pitch, and yaw) with variables that feed into a PID controller algorithm. This type of controller is very popular in large-scale industrial control systems. The process works by creating a control loop feedback that calculates an error value and attempts to correct said error to a desired result.

Luckily, APM comes with standard settings here that seem to work well for most installs right out of the box. If your aircraft is not flying as you would like, and you feel that you need to adjust the PIDs to obtain peak performance, we do recommend that you do some independent research on the topic. PID tuning can become pretty complex pretty quickly, and as such, falls outside of the scope of this book. There are also options to Auto-Tune your PID settings with APM, but we have seen mixed results with this process. Even if this is an option for you, it would still be a good idea to learn more about the theory of PID tuning so that you understand what is going on behind the scenes. If you are interested in learning more, we have a good starting point (*http://bit.ly/pid_tuning*) on the website.

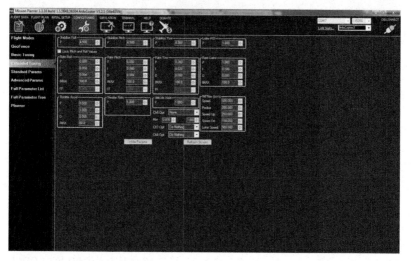

Figure 9-25. *Extended Tuning screen.*

Standard Params

The Standard Params page starts to show you the extreme flexibility of the APM platform (see Figure 9-26). The screen shows a long list of parameters that can be set to help customize or fine-tune your aircraft for your specific needs. Although this list is much too long to discuss here, we certainly invite you to explore it yourself; however, a word of warning: be careful what you change! If you don't understand what something does, look it up on the APM website before changing it and attempting to fly. A full list of the Standard and Advanced Parameters (*http://bit.ly/copter_parameters*) can be found on the Ardupilot website.

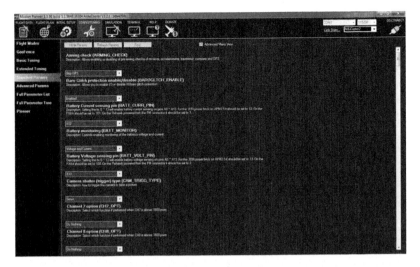

Figure 9-26. *Standard Params.*

Flight Data

For a good chunk of this book now, we have talked about telemetry data being shipped back to your ground station over the telemetry radios. The Flight Data screen (see Figure 9-27) is where we actually get to see most of that data. When you open the Flight Data tab you will notice a large map on the right side of the screen. This will give you a real-time update of where your aircraft is located (as long as you have a GPS fix).

On the left side of the application will be a column broken down into two sections top and bottom. The top portion contains an instrument that may look familiar to you. It is essentially an artificial horizon just like you would see in a full-scale airplane. The only difference here is that our horizon has some additional data overlaid on top that pertains directly to our type of aircraft. This instrument will display the attitude of the aircraft as it moves through space by moving our horizon line—separated by green and blue fields—as well as the aircraft heading, flight mode, GPS status, battery levels, and compass heading.

The space below the artificial horizon is used to display a number of data points that are returned from APM to our ground station. These data points can be very useful during flight because they help you gauge things like how fast your aircraft is flying, how far you have flown from your takeoff point, and what your current altitude is.

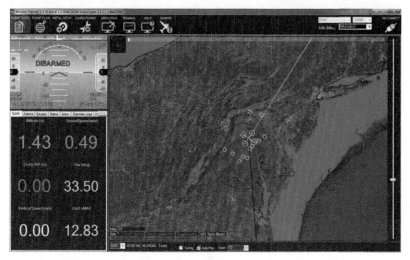

Figure 9-27. *Flight Data screen.*

Flight Plan

Finally, we will take a quick look at the Flight Plan screen, which enables you to plan your autonomous flights (see Figure 9-28). At first glance, it looks very similar to our Flight Data screen because it has the same type of map, but this map has a vast amount of additional functionality built into it.

Everything from an entire mission to a single waypoint can be planned from this screen. No matter what the complexity of your mission, they are all made up of a series of waypoints that direct the aircraft to a specific latitude, longitude, and altitude. You will also tell the aircraft what it should do once it reaches each waypoint—for example, you may want it to pause for 30 seconds or move directly on to the next way point—and how fast it should fly in order to get to that waypoint.

It is very important that you have a solid grasp on manual flight before attempting any type of autonomous flight. Every pilot should be able to manually take control and fly the aircraft home in the event of a GPS or sensor failure.

Figure 9-28. *Flight Plan screen.*

Autonomous Tablet Software

Although Mission Planner is a robust application that is more than capable of planning and executing autonomous flight plans, it's not always the easiest thing to use out in the field due to the fact that you need to run it on a laptop. Many people like to use smaller Android tablets to cut down on the amount of gear they need to pack for their flights. Tablets can even be mounted directly to the 2.4 GHz radio transmitter in an effort to keep all control systems in one neat location. If you are interested in exploring tablets for field work, download the 3DR Android application called Tower. This is the newest version of the long-used Droid Planner application. If you want to learn more about tablet ground stations (*http://bit.ly/ground_stations*), check out the details on our website.

Preparing for Your Maiden Flight

Technically, at this point, you are done with your build. All of the components are installed, your flight controller is set up, and you are just about ready to test it out and see how well you can fly! However, we do want to offer you just a few quick bits of advice first. If you are new to RC flight, we recommend that you finish up the remaining chapters to get a good understanding of how to fly in a safe manner. After you've done so, you can proceed with the following steps:

1. Leave your propellers off and double-check that all your motors are spinning in the right direction. If you remember from earlier in the book, we plugged the bullet connectors from our ESCs into the motors without knowing if we had the correct direction yet. Now is the time to double-check this. Arm your motors (default is left stick down and to the left for three seconds) and check each motor one at at time. If you find it is spinning in the wrong direction, change any two of the three bullet connections to reverse the direction.

2. With the props still off and the motors spinning in the correct direction, check to see if the APM is autoleveling correctly. You can determine if this is right by spinning up the motors slowly while holding the quad and gently dipping the frame in one direction. If the APM is working as expected, it should automatically increase the speed of the motors that are dropping closer to the ground. If you find this is not correct, double-check that you have all your ESCs plugged into the right output ports on the APM.

3. If everything is working as expected so far, feel free to put your props on and try to get this bird in the air! Take your time and start off slow. Spin the props up gently and see if you can make the quad just barely come off the ground. Give it some mild stick commands while doing this. Confirm that everything works as it should, such as a Roll Left command actually causing the aircraft to fly to the left. If all seems OK, gently give it a little more throttle until you are a foot or two in the air. Avoid staying too close to ground, where you will

get caught up in ground effect, which can cause you trouble. Ground effect occurs when air pushed through your props hits the ground and creates a disturbance that can push your aircraft around a little bit.

4. Take your time and practice! This isn't a race and there are no prizes for getting in the air immediately. If you take your time now and get the feeling of your aircraft down, you will be much more successful moving forward.

10/Safe and Responsible Flight

Above All Is Safety

There's a reason we hear the term *safety first* so often in our daily lives: doing what you can to protect the well-being of yourself and others is more important than any piece of equipment or task to be completed. Building and flying your hand-built small UAV safely and responsibly will ensure your continued ability to take part in this endeavor for a long time.

Those who persevere in building and flying small UAVs understand the importance of taking safety seriously. By doing so, they are respecting and representing the model aircraft and UAV community as a whole. With the concepts in this chapter in mind, taking part in aerial robotics is an outstanding way to learn or gain discipline. You must dedicate yourself to being mindful of consistently doing all that is needed to best produce successful and safe results in all possible ways.

Training and Education

A big part of staying safe while exploring multirotors is learning as much as possible about their technology, and its potential risks. New advances in small UAV technology in the areas of autonomous control and safety are being developed all the time, and staying up to date on these developments can keep your flights safe. However, we do not recommend relying on automated flight control modes. GPS locking and stabilization mode are fantastic features, but should any equipment fail to send or receive commands, a responsible operator knows how to take over manual flight of their aircraft. There's no substitute for building your own level of content knowledge, experience, and practice time "on the sticks."

Use flight simulator software (*http://bit.ly/simula tors*) to build muscle memory and instinctual flight control ability.

Resources

It is your responsibility to build and fly UAVs safely and responsibly, but you are not alone. If you are uncertain about any part of the design, build, or flight of your craft, it is OK to ask for help. There are many resources and online forums available to small UAV developers, where you can get advice and information on the best practices for UAV safety and technology.

At the end of 2014, three such groups worked closely with the US Federal Aviation Agency (FAA) to agree on policy statements and best practices for safely flying small UAVs. These groups, the Association for Unmanned Vehicles International (AUVSI), the Academy of Model Aeronautics (AMA), and the Small Unmanned Aerial Vehicles Coalition (SUAVC) compiled a concise list of the dos and don'ts of piloting small UAVs that they call "The Rules of the Air." Their suggested safety guidelines are a fine place to start, but we recommend that you actively and regularly look into the rules for yourself as they continue to evolve. We have summarized their list in the bullet points below, and have elaborated on them with some additional ideas for knowledgeable, responsible small UAV flight.

Important Links

Important links for the main SUAS resource groups include:

- AUVSI (*http://www.auvsi.org/home*)
- AMA (*http://www.modelaircraft.org/*)
- SUAVC (*http://www.smalluavcoalition.org/*)
- FAA (*http://bit.ly/faa_uas_dos_donts*)

In addition, check out our list of comprehensive safety procedures (*http://bit.ly/uav_safety*).

The AMA has a one-page PDF of best practices for flying RC aircraft (*https://www.modelaircraft.org/files/105.pdf*) as well.

Key Flight Safety Rules

Safe and responsible flight should always be your number one concern when preparing to take to the air. Become familiar with the following items, and you will be well on your way to sharing the skies with those around you:

- Always fly below 400 feet. Full-scale aircraft fly above 500 feet. This creates a 100-foot buffer zone in the airspace between manned and unmanned aircraft.

- Fly your aircraft within line of sight (LOS). This means you are constantly able to see your aircraft while operating it. From our experience, looking at your aircraft in the same direction as the sun can make visibility difficult. Think about the time of day and desired direction where you will be stationed. Colored landing gear or LED lights help identify the front and back of the multirotor, which assists in maintaining your orientation. See the note below this section on safely flying using FPV goggles.

- Join a local club for UAV or model aircraft enthusiasts. If that is not possible in your area, create your own MeetUp group to discuss safe and responsible developments on best practices of UAV flight.

- Never fly within five miles of any airport, within three miles of large stadiums between an hour before and after events, and not at all in national parks or military bases. Check out this interactive map (*https://www.mapbox.com/drone/no-fly/*) that identifies the no-fly zones.

- Take a flight lesson. This will help to reinforce the principles of flight, and it will let you experience navigating the airspace from a full-scale pilot's point of view. You may also be able to find a local course on operating small unmanned aircraft.

- Always inspect your equipment to make sure every component is in proper working order prior to every flight.

- Do it for fun! Don't fly for commercial purposes without authorization from the FAA. You should have years of experience before this should be a concern anyway. Insurance on your aircraft is not required for recreational use, but having it is a great idea. If you join the AMA flying club, limited insurance coverage is included in the membership. Insurance is a requirement for commercial use.

- Never fly recklessly. Not only is it dangerous, and disrespectful to the people and property in the area, but you may be issued a citation and a hefty fine. Fly safely.

First-Person View

Many multirotor enthusiasts love to fly first-person view (FPV) using video goggles that give the pilot the sense of sitting in the cockpit. A small camera mounted to the front of the UAV allows for a real-time view from the perspective of the drone. FPV flight is growing fast in popularity thanks to organized, competitive minidrone racing. The April/May 2015 issue (*http://make zine.com/make-44/*) of *Make:* magazine is full of great FPV features and information. Here are some tips:

- Bring a friend with you to act as a spotter. You need someone to be your second set of eyes to keep the copter within line of sight and alert you to anything that may interfere with your flight path.

- Clearly communicate what video channel/frequency you are using with any other FPV flyers nearby to avoid interference in your reception and visibility.

- Go to a wide-open, secluded location away from people, property, roads, and power lines.

- Avoid bringing children or pets out to the FPV flying field area because they may unknowingly enter your flight path.

- Follow all of the other safety guidelines outlined in the previous section.

Where and When to Fly

Planning is a critical step in safe, successful flight. There is a saying: fail to plan; plan to fail. There are many things to consider before your maiden flight, and every takeoff thereafter. We use a preflight checklist to be certain that we have gone over all functions necessary for our planned mission. You can use your favorite productivity app to generate your own preflight checklist. Many are available in the online app stores. The to-do list app we have enjoyed using is Wunderlist. It syncs easily across multiple platforms. It's also free, which is nice. See the next section for specific points of information you may want to include in your preflight checklist.

We also like to use the 3D feature in Apple Maps to plan our missions. You can view a perspective very close to what you are trying to achieve while in flight. Time of day and lighting features may also be important to you, if for no other reason than to ensure you're not looking directly into the sun while visually tracking your UAV.

The best way to identify potential risks is in person. We like to visit a location in advance to see if there is enough space, and if the location presents any possible hazards. Develop a plan to best mitigate risks to such a degree that is tolerable to all involved with the flight area. You likely need to get a permit or at least permission to fly in an area, even if it appears to be an empty field. If you own your own very large yard, you have the ideal setup for stress-free, regular practice space, as long as it is five or more nautical miles from any major airport. Finally, US national parks are (at the time of this writing) considered no-fly zones. We hope this changes at some point in the future. At the very least, we would hope to see procedures put in place for aerial photographers to obtain a permit to access national parks under the rules of the National Park Service.

Preflight Checklist and Flight Log Information

Just as in full-size aircraft, preflight checklists and flight logs are essential to maintaining a safe aircraft and flying environment:

- Date and time.
- Location and safe takeoff/landing area established.
- Operator and any flight team members such as spotters or camera operators.
- All wiring and hardware connections are secure.
- Aircraft, radio and channel, flight modes/settings
- Propellers and batteries used. We like to label and track each battery's usage.
- GPS: Number of satellites locked in.
- Weather, sun direction, wind direction and speed. Maximum safe wind speeds depend on the weight and design of your aircraft. The Little Dipper is very light at 2 lbs, so it is best to fly it below 10 miles per hour. Heavier multirotors can handle stronger wind. Also, avoid precipitation. Water and electronics do not mix well.
- Purpose/subject, mission, and contact person.
- Potential dangers and plan for handling each.
- Elevation/speed reached.
- Payload secure—best to start off payload-free.
- Camera settings and memory card with available space.
- Flight length and observations—did anything irregular occur regarding the equipment or experience?

 Never fly over crowds or traffic.

Try your absolute best to fly away from anyone, but avoiding every person can sometimes be difficult. If any spectators are present during a flight, establish a safe takeoff and landing zone (see Figure 10-1). We keep a minimum of 30 feet between our UAV and any person or thing. We like to use safety cones to mark our safe zone perimeter. You could also use an extra large tarp with cones weighing down the corners. A tarp would keep dirt out of your UAV's sensitive electronics too. Creating a physical line of safety using rope, poles, paint, or field-lining powder is a great way to keep spectators at bay. If you have a spotter with you, have her talk to anyone nearby to let them know what you are doing, to direct spectators to stay away from the flight zone, and to always be aware of where the UAV is. Remember, flying near large stadiums with crowds in the stands is prohibited by the FAA an hour before and an hour after an event is scheduled (as well as during the event itself, of course).

Figure 10-1. *Always keep a safety zone around your aircraft during takeoff and landing.*

Actively look out for any cables in the flight area. Power lines in the air, and even loose cables on the ground, can be very dangerous while flying. Outstretched tree branches, light poles, and architectural features can also pose a threat. Keeping your distance and avoiding them entirely is the only way to completely avoid the risk associated with these types of things.

Aircraft Inspection

Making assumptions can lead to making a big mistake. Never assume that everything on your multirotor is working great simply because you saw it fly flawlessly yesterday. Always do a systems check before each flight. Testing is certainly a constant in this hobby. A more thorough inspection with regular routine maintenance, along with the accompanying documentation, is a good idea as well.

Test the motors and settings without the propellers. Lastly, add the propellers, stand back, and do a prop-directional test. If you notice anything wrong, or had to repair/alter anything, document it on the spot.

Review battery procedures (refer back to Chapter 5). Their age, how they are handled, and stored, and whether they are intact or have unused charge can all have an effect on Li-Po batteries. Take it upon yourself to know how to use, store, and dispose of them responsibly. Locate your local drop-off center able to accept used Li-Po batteries. Understand that Li-Po battery usage can be dangerous and requires your disciplined attention. You should have a fire extinguisher on hand just in case.

Flight and Maintenance Logs

Much of the same information from the preflight checklist can be used to generate your flight log. It is important to document each of your flights. This helps you improve your UAV build and flight performance, by recognizing patterns in data you collect. An added benefit of keeping a flight log is that if you are able to show consistent proactive effort on your part to always fly with safety as a priority, you are much better off should your motives while flying ever be questioned.

Besides a flight log, we also recommend keeping a separate maintenance log for your UAV build(s). This may be as simple as using paper and a binder, or a detailed file kept digitally in whatever mobile electronic device you prefer. Record each repair or improvement you make, and when you did it. Answer the questions, "What caused the problem?" and "Why was the repair or replacement necessary?" Complete any relevant testing following the completion of the work, and record the observations. Of course, using quality components helps cut down on head-scratching, and repair and documentation time. For example, metal versus plastic mechanisms will naturally be more durable and hold up longer.

What we said back in Chapter 2 still holds: record your decision-making process throughout your drone build, especially if it is your first aerial robotics experience. Describe what factors led you to make each component selection throughout your build. Compare parts or brands you considered, and what the result of each decision turned out to be. It may seem like an excessive exercise, but later on you will be glad to have even half of such information on hand.

Laws and Regulations

In many countries, hobbyists have enjoyed model flight for nearly 100 years. Throughout the 20th century, the rules of common sense guided the model aircraft community as it typically policed itself with little to no incident. Recently, the capabilities of these aircraft have increased tremendously, and this has attracted the interest of lawmakers, especially in the United States. We strongly advise anyone just getting into the hobby of model aircraft to check for local and national laws regarding the technology. This area is moving at such a rapid rate that anything we write will likely be out of date by the time this book hits the shelves. If you live in the United States, the FAA or AMA websites are the best places to start.

Laws by Country

This list of sites from a few of the countries that have enacted comprehensive policies for small UAVs can help you get started in finding the laws that apply to your area (the distinction between commercial and noncommercial applications is the main common factor around the world to determine which set of rules you should follow):

United States
Federal Aviation Administration (FAA) (*https://www.faa.gov/uas/*)

Academy of Model Aeronautics (AMA) (*http://www.modelaircraft.org/*)

Know Before You Fly (*http://knowbeforeyoufly.org/*)

Canada
Canadian Aviation Administration (CAA) (*http://bit.ly/caa_recreational_uavs*)

United Kingdom
Civil Aviation Authority (CAA) (*http://caa.co.uk/uas*)

Australia
Civil Aviation Safety Authority (CASA) (*http://www.casa.gov.au/*)

Germany
Luftfahrt-Bundesamt (LBA), or Federal Aviation Office (*http://www.lba.de/*)

11/Real-World Applications

Beneficial Drones

When used safely and responsibly, small UAVs can be flown for a wide range of applications. When we speak to students, they constantly come up with creative new uses for the technology. UAVs can greatly benefit industries, people, and our planet. Many universities now offer majors in unmanned aerial systems. The first, the University of North Dakota, began its program in 2009, and Kansas State University was the second US college to offer a UAS major. The need for UAVs in agriculture is what pushed schools in the Midwest to lead the way.

Here, we will explore what we feel are the top real-world beneficial applications for UAV use. This list is just the beginning of how drones are helping save time, money, and lives.

There is an international competition called Drones for Good that challenges engineers to design UAVs that benefit our planet in a new way. First prize is $1 million.

Aerial Photography

The desire to create photography with a unique feel and perspective is a strong motivator that gets many into the hobby of UAVs (see Figures 11-1 and 11-2). Lightweight, durable action sports cameras have played a key role in the popularity of small drones. The ability to view larger areas and gain a clear sense of an environment has led to applications in the motion picture industry, sports broadcasting, and journalism. We started off using Terry's homemade UAVs for photographing landscapes and architecture in 2010.

UAV and camera technologies continue to progress rapidly. At this point, we prefer flying with a micro four thirds camera model—basically a hybrid between point-and-shoot and full size DSLR model. An important feature to have is a remote sensor that enables you to trigger the shutter from the ground. Another method is to set the camera to take pictures at regular intervals, say every five seconds or so. We have found that a two-person team works well to compose the best shots. The pilot is able to concentrate on operating the aircraft, while the camera operator uses a second controller for the gimbal and camera. A ground station monitor enables you to see the view from the onboard camera and refine your aerial position to capture the final composition.

Refer back to Chapter 8 for more information on UAV camera equipment.

Figure 11-1. *Francis Scott Key Monument, Baltimore, MD, taken with Sony NEX-5n.*

Figure 11-2. *Firemen's Carnival, Baltimore, MD, taken with Canon S95.*

Mapping and Surveying

Drones are proving to be an efficient tool that is revolutionizing the field of geographic information systems (GIS). The drone's payload can collect a range of data using cameras and sensors. These devices, used along with geo-referenced ground markers, produce highly accurate maps with a resolution up to five centimeters per pixel.

Two main types of software are involved in UAV mapping. The first type is for mission planning, and the second is for data processing. The UAV flies in a lawnmower pattern autonomously (see Figure 11-3). Several mission planning applications are available including free, open source apps. The mission is uploaded to the UAV in its entirety. While flying the mission, the payload points straight down and captures data at regular intervals, creating overlapping images of the area—at least 60% front to back and 40% side to side. Once the data is collected, it is processed by removing distortion and stitching images to create an ortho rectified mosaic. As with the navigation software, there is also a range of data processing solutions from locally installed applications to cloud-based services.

Figure 11-3. *The white line shows a UAV mapping flight path in a lawnmower pattern.*

Precision Agriculture

New technology is essential for keeping up with the nutritional needs of a growing population. With 2.2 million farms in the United States alone, precision agriculture will benefit immensely from small UAVs. Farmers, agronomists, and farm support service providers are all aware of this and are more tech savvy than you may think. Drones are less expensive and more accessible than regular airplanes. UAVs can also complete the tasks previously done with manned aircraft much faster with more precise results. Additionally, because farms are generally located in scarcely populated areas, privacy and safety issues regarding drones are much less of a concern.

There are two main areas in which UAVs assist farmers. The first is with GIS, and typically utilizes a fixed wing style of drone aircraft. They do not require as much power, which helps them stay in the air longer. By gathering data from infrared sensors and stitched image files, farmers are able to efficiently map the farm, estimate crop yields, assess plant health, identify weeds or diseased plants, record plant growth, and measure hydration levels. All of this information is then analyzed and used to make the best crop management decisions. Prescriptions for fertilizers and pesticides can then be selectively applied to individual plants in an improved field plan. This is where the second type of UAV aircraft comes in. Helicopters or multirotors capable of carrying a payload are used to spray various prescribed formulas, row by row, precisely onto only those crops that need it.

Successful UAV precision agriculture programs are found in Canada, Brazil, and Sweden, but the standout model is in Japan. The Japanese Agriculture Ministry commissioned Yamaha to develop an unmanned aerial system for agriculture in 1983. The Yamaha RMAX unmanned helicopter has been spraying crops since 1991. Today, the RMAX sprays 40% of Japan's rice fields, approximately 2.5 million acres. Their yield increase has been estimated as high as 30% while maintaining an impeccable safety record. Drones will indeed revolutionize farming.

Search and Rescue

Drones have assisted many times in the successful location of missing persons, whether the incident was due to a natural disaster, abduction, or simply getting disoriented and lost. A UAV equipped with a high-resolution camera or infrared sensor can search hundreds of acres in a matter of minutes. To cover the same area on the ground would take dozens or hundreds of volunteers hours or even days. The initial period of time from when a person is first reported missing is the most crucial, and mobilizing resources early can make all the difference in finding them alive.

Some examples of teams that are effectively designing or using drones for search and rescue at a fraction of the cost and risk include:

- Royal Canadian Mounted Police found a family that got lost for days while hiking in a forest near Topsai Lake outside Nova Scotia.
- New Zealand Coast Guard uses its drones to find people and boats lost at sea. Robolifeguard is a sUAV that has a shark-deterrent module.
- A Swiss group has developed a drone for tight, indoor missions called the Gimbal UAV. It is housed in a cage that can bounce off obstacles or roll across the ground.
- David Lesh, an amateur drone user, found Guillermo DeVenecia, an 82-year-old missing man who wandered into a soybean field. After days of searching using other methods, Lesh flew at an elevation of 200 feet and quickly spotted DeVenecia.
- Texas nonprofit EquuSearch gained widespread attention for filing a petition for review with the US Court of Appeals for the District of Columbia, with a complaint against the FAA for wrongly ordering a cease-and-desist in an email for using UAVs for search and rescue. The court found that the FAA's email was "not a formal cease-and-desist letter representing the agency's final conclusion." Equusearch's missions have contributed to finding many missing persons.

Infrastructure Inspection

Aging infrastructure and the rise of extreme weather can have disastrous effects on transportation, communication, and energy infrastructure. Inspection of the structures related to these industries is where small UAVs present many advantages over regular manned aircraft (see Figure 11-4). Drones are able to achieve tight close-ups and get into hard-to-reach areas. Safety is another clear advantage. Consider workers who must climb structures or use bucket trucks and cranes to suspend themselves from the sides of massive bridges and skyscrapers. These outdated methods are simply dangerous, time-consuming, and expensive. For the cost of one manned flight, a company can own and operate an entire UAV imaging system.

Being able to hold the UAV at a precise set of GPS coordinates, plus the ability to move an airborne camera in three axes, allows an operator to obtain an extremely clear view of any area technicians would need to get a detailed look at. A live feed allows the inspectors to see in real time what the drone sees. High-resolution files can then be stored and reviewed for making assessments and developing repair strategies.

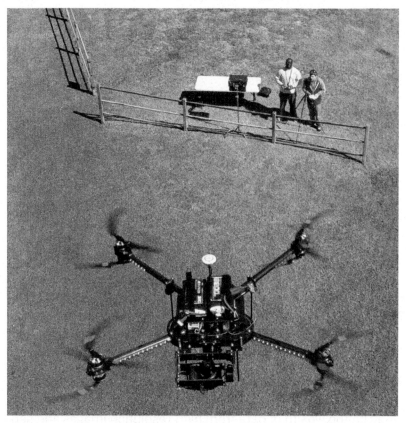

Figure 11-4. *Air-to-air image of the RC Rotors Inspection Drone with camera point straight upward; the X8 is piloted by Beresford Davis, and the photo is by Terry Kilby.*

Drones are a game-changer for inspecting numerous structures including:

- Bridges
- Power lines
- Cell towers
- Buildings
- Roofing
- Oil pipelines and rigs
- Nuclear plants (radiation measurement)
- Thermal analysis of buildings
- Gas refinery flare tips
- Water treatment facilities
- Wind turbines
- Smoke stacks
- Hydroelectric plants
- Ships
- Solar arrays
- Historical monuments
- Railways
- Highways (overpasses, tunnels)
- Large equipment (cranes, earth movers)

Conservation

Biologist and climatologists face many difficult challenges in their fight to save our planet's endangered species and their habitats. They must cover vast, arduous-to-navigate locations. Planning expeditions is expensive, laborious, and time consuming. Conservationists have perished from encountering armed poachers. Many over the years have died flying in light aircraft at low altitudes for their observations. Out of dire necessity, scientists have welcomed the many advantages gained using sUAVs to efficiently complete their work. The speed, safety, and time and cost savings are wonderful. However, the greatest benefit may be the wealth of data from high-resolution images and sensors never before seen, much less gathered. It seems drones have given conservationists renewed hope.

12/Expanding Your Drone's Abilities

Congratulations! You have done it. You built your first drone, and you have learned a number of valuable lessons along the way. It is our hope that at least some of these lessons will resonate with you in a way that encourages experimentation in the future. This chapter gives you a few suggestions for modifications that you can make now that you've mastered the art of building drones.

Add a Camera and First-Person View

If you built the autonomous kit, it did not come with an FPV package or camera. Now that you are comfortable building these aircraft, it's a great time to add a live video feed to see what your drone sees in real time. To do this, you will need to add a video transmitter on the aircraft and a video receiver to a ground station with some type of monitor (goggles or small screen) to view the feed (see Figures 12-1 and 12-2).

FPV Installation

The drone model, the Little Dipper, that we used as the build example for this book is a 300-class quadcopter, which is small compared to other drone designs. For this reason, to add FPV to such a compact quad, we recommend that you remove another component (that's not vital for flight). If you have added everything shown in this book, there is already a good deal of electronics loaded onto your drone. For best flight performance, battery life, and overall stability, it is best to avoid trying to carry too much on your drone's frame. If FPV is a must for you, perhaps removing the telemetry radio is the right way to go.

Installing an FPV system can be an entire chapter of a book, and falls outside the scope of this one. However, if you are interested in the topic and would like to learn more, feel free to check out our online FPV page (*http://bit.ly/fpv_install*).

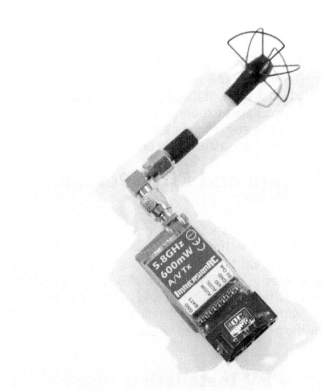

Figure 12-1. *5.8 GHz video transmitter with a right-angle adapter and right-handed Skew Planer antenna.*

Figure 12-2. *The Mobius action sports camera—a great option for an aircraft the size of the Little Dipper—is very popular due to its light weight and high performance.*

Collect More Data with Other Sensors

APM is able to incorporate a number of additional sensors. Because this is an open source platform, you can even alter the firmware to react to certain sensor input any way you like. Perhaps you would like to use a wind speed sensor to slow down forward flight once a certain speed has been reached, or perhaps a proximity sensor to detect when you are a certain distance from the ground. This is all possible with a little research and some clever code. The fantastic DIY drone community that has sprung up online can serve as a very valuable resource for such projects.

Great Web Resources

A great way to get started writing custom firmware or incorporating custom sensors is to find others who have already experimented in this area, and build on what they have done. DIY Drones (*http://diydrones.com/*) serves as a great community to link up with other developers.

If you are interested in writing custom firmware or software for your newly built drone, look into Drone Kit (*http://dronekit.io/*); it's a new middle layer that sits on top of the original code base and allows you to easily write custom code in the Python language.

Altering Speed: Ready, Set, Race!

One of the first mods that we are always asked about when it comes to drones is how to make them faster. As you have already learned in this book, getting the desired performance requires a fine balancing act among all the components and the design. That same theory applies to speed as well. The major things to consider when trying to increase the speed of your quadcopter include:

Weight
> As with anything that is built for speed, weight is the very first thing you should look at when trying to improve performance. Anywhere that you can shave a few grams will allow you to see some increase in your aircraft's capabilities. The more you can shave off, the better you will be in the long run, even if it's just a few grams here and there. Consider removing any parts that aren't mission-critical. Do you really need that telemetry radio just to fly fast? What about the GPS? Maybe you want to replace the autopilot altogether with something that is much lighter. These are all things to consider when trying to reduce weight.

Prop size and pitch

As we learned earlier, the props are much like the tires on a car and can greatly affect the real-world performance of our aircraft. If you want to increase the speed of your quad, try experimenting with different props. The higher the pitch, the more air you will move through the prop, which generally equals a higher rate of speed. Be warned, though: there is such a thing as too much pitch. You may notice a decrease in stability of your aircraft at a certain point. Props are a fairly inexpensive part of this equation, so we will often buy several different types to play with and find out exactly how they perform on our own gear rather than taking advice online. Something that doesn't work for someone else may work great for you, and vice versa. On the Little Dipper, a 6-inch prop is the largest that you can fit, so try experimenting with different pitched props of that size.

Battery voltage

In Chapter 3, we talked about how you can calculate the RPM of your motors by multiplying the battery voltage against the KV rating of your motor. In those equations, we only discussed a three-cell battery that has a voltage of 12.6 with a full charge. Each volt, as we already learned, has about 4.2 volts of power when fully charged. What we didn't explore was the vast array of battery types out there. What if you changed your three-cell battery out for a four-cell? Now you are talking about a maximum voltage of 16.8 and an increased motor speed of 9,660 RPM! That can make a huge difference when building a racing quad!

 Increasing Battery Voltage

Increasing the operating voltage can have an effect on your other components. Make sure that everything in the power train is rated to work properly with the voltage you plan to use. The ESCs, motors, and autopilot all need to be rated for the voltage that you are looking for.

Maybe the winning combination comes in the form of a smaller capacity (for lighter weight) higher-voltage battery (for higher RPM) combined with shorter props (for quicker response) that have a higher pitch (to move more air). Experimenting with all these different options is what we consider to be the fun part of building these aircraft.

Increasing Flight Times

The general rule of thumb for increasing flight time is to go with a longer prop that has a lower pitch. In the case of the Little Dipper kit, we have already bundled it with the 6-inch (largest that can fit on the frame) prop at a pitch of either 4.5 or 3. There may not be a lot of upgrade room left, but it is possible to experiment and see what you can do. At the time of writing this chapter, 6 × 2 props are just starting to hit the market, and they are a good possibility for longer flight times. Of course, decreasing the weight of the aircraft will also gain you a little more flight time, so look at that as another option to improve your air time.

Troubleshooting in the Drone Community

Our final recommendation in this book is that you come to terms with the fact that as long as drones are in your life, repairs and problem solving are as well.

Troubleshooting issues that arise and affect flight performance are a normal part of being a small UAV enthusiast. You will be completing routine maintenance and systems checks before each flight to minimize unfortunate events while flying, but repairs are inevitable. Luckily, you are not alone. As mentioned in Chapter 1, there is an entire, well-established online drone community that we encourage you to get involved with.

Helpful and popular online forums include:

- DIY Drones (*http://diydrones.com/*) is a really great resource if you are using APM components.
- MultiRotorForums.com is a fantastic resource filled with experienced pilots and beginners alike.

- YouTube has a multitude of drone tutorial videos (*http://bit.ly/drone_tutorial*). Check out the FliteTest crew (*http://bit.ly/flitetest_vids*).
- UAV component manufacturers may also provide helpful information. DJI (*http://www.dji.com/*) has incredible print and online instructional information.
- There are a multitude of UAV Facebook groups for specific frame designs and manufacturers, including a wonderful group for female builders and pilots called the Amelia Droneharts.
- Maker Media (*http://makermedia.com/*) has a phenomenal lineup of instructional publications including *Make: magazine* and the entire series of *Getting Started* books.
- We also invite you to share photos of your drone builds to our website's gallery. We would love to see how you took the concepts in this book and ran with them. Use the incoming-only email address to send in pics of your multicopter builds: *userbuilds@gettingstartedwithdrones.com*

Thank You!

Speaking of the online drone community, we would like to say thank you to everyone from DIYDrones and MultiRotorForums who has shared any tidbit of information they learned the hard way, and was generous enough to share with us over the last almost six years.

We must also thank a special group, our fellow flight-obsessed bunch who started out with us in a local Baltimore MeetUp group. They never fail to graciously contribute their time and talents toward our endeavors. They've kept us laughing too through all of the shaky takeoffs and hard landings.

We appreciate you: Brian Kraus, Chris Meaney, Tom Minnick, Ian Wollcock, Beresford Davis, and Elliot Greenwald (who was our awesome hand model)!

Finally, many warm thanks to you, the reader, who have made the investment in time, money, and effort to challenge yourself to get started with drones. Hope to see your builds in the gallery! We wish for you many years of learning and happy flying!

Index

throttle, 10
thrust, 9, 43, 44
thrust vectoring, 11, 15
tools, 22, 52, 83, 110, 116, 126, 131
total lift, 48
Tower software, 117, 158
tractor props, 41
training and education, 161
transmitter (radio, remote control),
    107-108
    (see also receiver)
    calibrating, 142
    communication protocols for,
        109
    frequency bands for, 108
    Mode 1 or Mode 2, 10, 108
    stick mapping for, 9-12
transmitter (video), 125-126, 179
tricopters, 17
troubleshooting, 184-185
tuning APM, 153-154

U

UAVs (unmanned aerial vehicles),
    vii, 1
    (see also drones)
universities offering majors in
    UAVs, 171

user community, 2-3, 184-185

V

vector thrusting, 11, 15
vibration isolation plate for camera,
    127, 128
video transmitter and receiver,
    125-126, 179
volts, 49

W

waypoint navigation flight mode,
    93, 144
weight of drone, 7, 182
where and when to fly, 165
wind speed sensor, 181-182
wire color standards, 86

X

X frame type, 136-138
X8 aircraft, 18

Y

Y6 aircraft, 17
yaw (rudder), 11

About the Authors

Terry Kilby has worn many different hats over the years. There might not be anything tech-oriented that he hasn't at least tinkered with in the past. Photographer, author, product designer, software engineer, entrepreneur—these are all titles he has held at various points during his career. In 2010, he became interested in multirotor aircraft as a way to capture new and interesting photos. He was immediately drawn into aerial robotics, because it touches on a wide range of skills he enjoys using.

Belinda Kilby's dedication to visual art throughout her school career helped her win many merit-based scholarships, which she used to attend Maryland Institute College of Art. After becoming a young mother, she transferred to Salisbury University and University of Maryland Eastern Shore to earn degrees in art and education. Art history was another favorite subject area for Belinda. She had the privilege of seeing many great works in museums when traveling while growing up. Travel and movement emerged as symbolic themes in her visual work. Belinda went on to teach art for Baltimore City Public Schools for 10 years. Speaking to groups regularly has prepared her to present information about the benefits of building and operating small UAVs in a safe and responsible manner. With her education background, it was natural for her to write an aerial robotics curriculum.

Colophon

The cover and body font is Benton Sans, and the heading font is Serifa.

CPSIA information can be obtained at www.ICGtesting.com
Printed in the USA
BVOW11s1057091015

421585BV00001B/1/P